MENTAL BIOLOGY

MENTAL BIOLOGY

THE NEW SCIENCE OF HOW THE BRAIN AND MIND RELATE

W. R. KLEMM

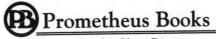 Prometheus Books

59 John Glenn Drive
Amherst, New York 14228

Published 2014 by Prometheus Books

Cover image © Media Bakery
Cover design by Nicole Sommer-Lecht

All interior images were created by the author, unless otherwise noted.

Inquiries should be addressed to
Prometheus Books
59 John Glenn Drive
Amherst, New York 14228
VOICE: 716–691–0133
FAX: 716–691–0137
WWW.PROMETHEUSBOOKS.COM

18 17 16 15 14   5 4 3 2 1

Library of Congress Cataloging-in-Publication Data

Klemm, W. R. (William Robert), 1934-
 Mental biology : the new science of how the brain and mind relate / by W.R. Klemm.
    pages cm
 Includes bibliographical references and index.
 ISBN 978-1-61614-944-4 (pbk.)
 ISBN 978-1-61614-945-1 (ebook)
 1. Brain. 2. Mind and body. 3. Consciousness. I. Title.

QP376.K4953 2014
612.8'2—dc23

2013040030

Printed in the United States of America

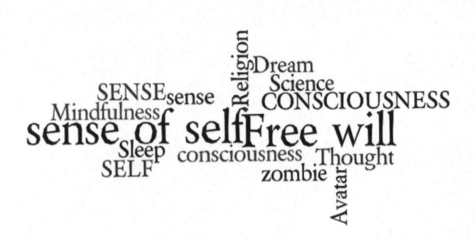

CONTENTS

We make up our mind. We say what's on our mind.
We change our mind.

We say we, or others, are mindful, mindless, mind numb.
We sometimes say we have lost our mind.
We say, "We are what we make of ourselves."
But who is the "We" doing these things?
This book seeks to understand the human mind
    so we can stop acting so much like zombies
and make our brains serve us better.

ACKNOWLEDGMENTS

This book would not have been created without my teachers. My first debt of thanks goes to David MacPherson, my deceased high school vocational agriculture teacher. He compensated for my lack of high school science education (I only had one year of chemistry, and sometimes I had to explain things to my football-coach chemistry teacher). But MacPherson enabled me to learn things I could not get out of a biology or physics book. I learned to value life and living things, particularly the mental life of the various animals I raised (chickens, ducks, geese, goats, pigs, and cows).

Next I thank my science professors at the Universities of Tennessee, Auburn, and Notre Dame. Before I went off to graduate school at Notre Dame to learn how to be a scientist, I was a Preventive Medicine Officer at a US Air Force hospital. Being stuck in the middle of nowhere, I spent a lot of time reading research journals in the library. One day a bolt of revelation struck me: "I understand this stuff. I can do this!" My professors had prepared me well to learn from the essential source in science, the writings of practicing scientists.

Finally, I want to thank my agent Rita Rosenkranz for connecting me to the staff at Prometheus Books, who are real pros at what they do. Special thanks go to my editors Steven L. Mitchell and Brian McMahon.

PREFACE

Most of what we know about the brain has been discovered in the last hundred years, especially after World War II. I was lucky enough to come of age in this golden age of brain research. When I entered my formal studies of neuroscience in 1960, there was no neuroscience as such. Then, brain research was conducted by scientists working in separate, isolated disciplines, each of which focused more or less on one facet of brain function. Working on some aspect of brain function were biochemists, cell biologists, anatomists, physiologists, pharmacologists, pathologists, psychologists, animal behaviorists, and others who had their own separate scholarly societies and journals. As a critical mass of discoveries about the brain emerged, these people realized they needed to stop working in isolation. I entered this environment as a physiologist interested in brain function. Realizing this need for integration, I became a founding member of the new Society for Neuroscience in 1969. Beginning with a nucleus of five hundred scientists from multiple disciplines, the society has now grown to over forty thousand members. Annual meetings now have about forty thousand attendees.

Brain research is hot. In 1990, President George H. W. Bush declared the beginning of a "Decade of the Brain." President Obama in 2013 proposed a massive new program for brain research. At present, the field seems dominated by a fascination with brain imaging, where color-coded pictures reveal which parts of the brain are metabolically active at any one point in time. While such imaging is revealing useful things, it can also be highly misleading. I do not envision the future brain research to be in generating more pretty pictures of brain scans. At the other extreme of brain research, there is a focus on reductionism, now coalescing as the sub-discipline of molecular neurobiology, which I regard as learning more and more about less and less. Interestingly, I heard this same complaint from the pioneering embryologist Paul Weiss some forty years ago. While molecular biology has been and is important, it is not likely to be central to future discoveries about what it means to be fully human in terms of conscious mind.

The thrust of my book is about how brains work, when and how brains create consciousness, what consciousness is, what it does, how it does it, how those actions program and change the brain itself, and how each of us can use our consciousness to enrich our lives. I view consciousness from the perspective of womb to tomb; that is, I suggest that rudimentary signs of consciousness begin to emerge prior to birth, and I speculate about the possibility that life might not end in the tomb.

Translating the language of modern neuroscience, I try to show the truth of René Descartes' axiom: "I think, therefore I am." My corollary, which he did not state, is:

I will become what I think.

While Descartes' views have been accepted for centuries, there is no single-source explanation for why they may be true. In fact, many scholars are now saying his views are wrong[1] (see chapter 4).

I focus on the "I" that thinks itself into existence. This "I" is the mind, both unconscious and conscious, which I consider a "being," just like the body in which mind is embedded. This book explains how the human mind comes from the brain and is not some "Ghost in the machine," as they used to say in Descartes' day. I think brain function can be explained, to the extent that it can be explained, by understanding how unconscious mind works. I view *conscious* mind as a neural-activity-based avatar, created by the brain to act in the best interests of both the body and the brain. I will present evidence to show that this avatar acts, and sometimes even acts on its own with creativity and freedom of will. Many scholars today dispute both ideas.

Some of this material is a general-audience explanation of my scholarly publications during the last three years, which include peer-reviewed papers on the material basis of consciousness, free will, and why we have dream sleep. Some of my views on the biology of sense-of-self ideas appeared in a recent invited chapter in the book *Consciousness: States, Mechanisms, and Disorders* (Nova Science). My ideas on the neurobiology of human agency (how the mind gets the body to do things) are explained in the forthcoming volume 12 of the series *Annals of Theoretical Psychology* (Springer International). This present book aims to integrate these ideas—in accessible language—in one place and in the context of the larger existential issues of human experience.

Some special features of the book are:

- my ideas on how a sense of self is created in the womb;
- my new theory on how consciousness is created and why evolution favored the appearance of consciousness—I try to explain the human mind in terms of the physical nature of thoughts and how they are represented and processed;
- my explanation of the differences between nonconscious, unconscious, and conscious minds;
- why I think the brain has to learn how to be conscious;
- my challenge to the common interpretation of research that "free will" is an illusion;
- my explanation of the errors of scholars who think consciousness is only an "observer" that cannot do anything;
- my new theory for why dream sleep occurs in mammals and in humans in particular,
- evidence regarding how the human brain generates intent, identifies context, recalls relevant memories, evaluates choice options, makes decisions, as well as how it constructs, plans, and implements decisions; and
- my explanation of brain programming by the conscious mind and why what we think and do determines what we are and what we can become.

I conclude by showing that there is much that science has *not* discovered, some of which may be actual physical realities that could contribute to our understanding of mind. This includes things scientists know about but do not fully understand, such as relativistic space-time continuum, quantum mechanics, dark matter, dark energy, string theory, and parallel universes. What scientists *do* know makes it clear that life is a unique and seemingly magical category of creation, far more than the sum of its chemistry and physics.

This book is not only about the brain's odyssey from womb to tomb, but also about my own journey of understanding as brain research has progressed during my professional lifetime. Such understanding is important to me because I have been a small part of the process. It should also be important to you because your brain has its own odyssey and you have considerable control over that journey.

CHAPTER 1
IN THE BEGINNING

Pond scum is beautiful and elegant. At least scientists who study the microbes of pond scum think it is. What is magical about these microbes is that they are *alive*. Even though scientists can describe most of what such microbes do, and can explain them in terms of the laws of chemistry and physics, we should all marvel that anything that originates in the water of muddy ponds can be so intricate, so complex, so alive. Infinitely more astonishing and marvelous is the evidence that human intelligence, creativity, and spirituality came from such primitive existence.

The specialness of life is magnified by the very real possibility that it only exists on this one obscure planet in a vast universe of billions upon billions of galaxies. There are no signs of it anywhere else in the universe. Despite decades of searching for extracellular life with radio astronomy and rocket-launched probes, no one has found any.

The closest astronomers have come to finding life elsewhere is evidence that Mars may have once had water. Mars may have had its version of pond scum a few billion years ago, but clearly it cannot support life now. Life, as we know it at least, must have some kind of solvent, and water is the best-suited solvent. There are, of course planets that have frozen water, but the atoms and molecules used by living systems have to move around, and they can't do that in a frozen solvent. Life seems nearly impossible to come by and hard to sustain.

The key to life is temperature. Life requires a temperature that obeys the Goldilocks principle; that is, the temperature has to be just right, neither too hot nor too cold. If it is too hot, the solvent boils away from the planet. If it is too cold, the solvent is frozen. A Goldilocks temperature is rarely met in the universe. In our solar system, only Earth has a clearly life-compatible temperature. Searches for such planets in other solar systems have been largely unrewarding. Recently, however, a report was published on five such planets existing in a galaxy "far, far away," about fifteen hundred light years. Even so, these planets may not have water or other suitable solvent. There is no known way at present to know if there is life there. Even if these planets do support life, it is a huge leap for any planet to get from simple pond scum to biological complexity equivalent to that of humans.

I went to college for nine years, most of which was spent learning about living things. I now think of life as a miracle, qualitatively different from the limited world of chemistry and physics. Science cannot fully describe the difference between living and dead. We say that living things grow and reproduce, but dead crystals can do that. However, life evolves.

Even now we are receiving the light of stars being born, and the light of stars that are dying or exploding to create comets, planets, and the like. Theorists suggest that "multiverses" are evolving, being created by other big bangs. String theorists have reasons to postulate "miniverses" that operate on different scales, physical laws, and dimensions than the one our avatars live in. General relativity equations suggest the possibility of "worm holes" that allow rapid transit through the fabric of space-time to other universes or within different regions of our own universe.

While such physical evolution is profound indeed, evolution of life forms is even more awesome. Living things grow and reproduce in a quite different way, for they create qualitatively new kinds of things. The most important product of evolution is the progression from undifferentiated nerve cell nets in creatures like jellyfish to the human brain, which has a conscious sense of itself that thoroughly engages its inner and outer world.

Life is delicate. Living things die. Yet life continues to create new chemical systems, proteins, structures, genes, and species. Life continues as if some hidden force is driving it to beat the odds of the second law of thermodynamics, which predicts that all things must degrade to disorder and chaos. Though what I am about to say is not scientific, I think that sentient animals and humans do have some mysterious life force that science has so far not identified. Interacting with animals for over fifty years convinces me that they are not just blobs of reproducing protoplasm. Surely, most of us think about humans in a similar way. When I have been in the presence of a loved ones at the point of death, I have been overwhelmed by the feeling that some life force has been sucked out of them in some kind of instantaneous swoosh. I had a little of this sensation with my last pet dog. Some friends have said that they have had similar experiences. Have you?

This is the big picture view of why life is so special. The close-up view of the specialness of life is equally awesome. You see, living things, even the simplest forms of life, do certain things that nonliving matter can never do. Living things

- capture energy from the environment, especially in high-energy phosphate bonds, and use it to develop organized complexity and grow;
- use genes to create a code for that complexity;
- express the code in orderly, self-organizing, and predictable ways in response to the environment;
- reproduce; and
- evolve into different kinds.

Even the humblest microbe is this special. How much more so is human sentience? Does it matter if people understand and value the specialness, the sanctity, of life? Of course it does. Much of the evil in the world arises directly because so many people have little respect for life other than their own. From ravaging the environment to mistreating animals and fellow humans, evildoers neither restrain nor enlighten themselves with a belief in the sanctity of life.

FROM "BIG BANG" TO BIG IDEAS

In the beginning, scientists tell us, there was a "big bang" when an infinitely compressed speck of matter exploded, releasing its mass along with light that flooded the nascent universe. The evidence for a big bang is pretty solid. Astronomers observe the galaxies racing away from each other as if they all arose like shrapnel from a grenade. They even measure the rate of such movement. Surprisingly, these calculations show that the dispersal of galaxies is accelerating. Intuitively, the accelerating expansion of galaxies makes no sense because gravity should be holding them back and in any case would prevent acceleration. Shrapnel slows down with distance from the explosion source.

However, it appears that some as-yet undiscovered source of energy is driving galaxies apart from one another, like paint spots on the surface of a balloon that is being blown up. The observation of accelerating expansion suggests that the energy itself is increasing. The galaxies themselves seem to be holding together presumably because of their own gravity.

Nobody can prove scientifically that the universe or life itself has a purpose. Indeed, many people have concluded that there is no possibility of purpose. But I think there are just too many scientific observations suggestive of purposeful order for it to be dismissed out of hand. Nihilists point to the second law of thermodynamics then opine that all matter living and dead will degrade into disorder and chaos. Yet life goes on and continues to beat the odds against it. It is as if life is *supposed* to exist.

Scientists can develop pretty good explanations of the creation and evolution of the universe and even its living products from the laws of chemistry and physics. The task of science is to discover and understand how those laws apply to living things, particularly the mind.

VICTIM OF BIOLOGY AND CIRCUMSTANCE?

Most folks are not concerned with the cosmos. The concept is too mind-boggling, even for many astronomers. But people do want to know about purpose in their own life. One purpose, often unrecognized, is to deploy mental biology in the service of shaping brain development in ways that make life more meaningful and fulfilling. People don't often think about that because they take what they are—their genetic endowment and life circumstance—as a given.

Charles Darwin and his "bulldog," Thomas Huxley, perturbed the subsequent generation's biological understanding by marshaling the evidence for biological evolution. Since then, physicists and astronomers have made an even more compelling case that the universe itself is still evolving.

Opponents of biological evolution are troubled by two claims of science:

(1) that complex life forms, including humans, evolved from primitive ancestors, and

(2) that natural selection, not the hand of God, is the force that drives the evolution of species.

To rationalize their objections, evolution opponents dismiss evolution as "just" a theory. Unlike evolution of the universe, which we can see and measure with the tools of astronomy, biological evolution is harder to prove. The other problem is a lack of understanding and a dismissive attitude regarding scientific theory. Theory drives scientific advance, a point that people without a science education commonly do not appreciate. Scientific theories provide the conceptual framework for generating testable hypotheses. Over time, if enough of these hypotheses survive experimental testing, the underlying theory becomes progressively accepted as useful and valid. Otherwise, the theory is either modified to fit new evidence or discarded in favor of a theory that successfully explains new observations and evidence. So far, no *scientific* theory has been able to dislodge evolution from its lofty perch.

Another area of controversy in the life sciences relates to the relative roles of genetics and the environment. Confusion commonly afflicts politics. For example, early Communists glommed on to the discredited genetic theory of "inheritance of acquired characteristics." This theory holds that changing a person's attitude and behavior would somehow result in changes to his or her genes, which would allow for genetic transmission of the changed attitudes and behavior to his or her children. For this idea to be true, outside influences on the brain would have to change the genes not only in brains but also in the sex cells (sperm and egg cells). The idea was held in ancient times by Hippocrates and Aristotle, but it gained scholarly imprimatur with formal publication in 1809 by Jean-Baptiste Lamarck. In the 1930s, the Russian president of the Soviet Academy of Agricultural Sciences, Trofim Lysenko, applied the doctrine to Soviet agriculture with disastrous results. At the same time, Soviet political leaders extended the mistaken doctrine to inheritance of educational and social experiences; that is, changing human nature by government policy. They expected that indoctrinating the current generation in collectivism would genetically transfer collectivist attitudes and behavior to all future generations. Cuba, North Korea, and China showed that collectivism can be transferred culturally but not biologically.

In the United States, much political angst arises from disputes over whether more effective educational and social policies will succeed in lifting people out of poverty and dysfunctional behaviors. When I was a child, I often heard the axiom, "You can take the boy out of the country, but you can't take the country out of the boy." Today, the corresponding axiom would seem to be, "You can take the boy out of the ghetto, but you can't take the ghetto out of the boy." The reality is that you can take the country or ghetto out of the boy, but this won't transfer to his children by his genes.

What we are now discovering is that environment and experience affect the *expression* of genes. Whether or not genes are accessible for readout often depends on the environment. People have underestimated their capacity to sculpt their own brains, attitudes, and behavior by controlling experiences that affect gene expression. However, though people may control to some extent how their own genes are expressed, there won't be any biological transfer to their heirs. I explore this topic in chapter 4. Environmental and cultural influences do of course transfer, so one's heirs can be *taught* how to likewise exert control over how *their* genes are expressed.

Having the right chemicals in the right environment at the right time is believed by most scientists to be all that is needed for creating life and shaping the mental life of the individual. To them, life seems like a highly improbable occurrence.[1] But it did happen, and even more improbable, there may be a life force that sustains it.

Many scientists also think of the brain's conscious mind as an emergent property of brain function. Emergent properties follow the rule that the whole is greater than the sum of its parts. Another way of saying this is that the properties of the whole cannot be predicted from what you know about the properties of the contributing parts. Yet, paradoxically, most scientists believe that as they learn more and more about less and less, they will somehow explain the whole.

Emergent properties apply both to molecules in a primordial soup that generate simple living organisms and to the 100 billion or so neurons of a human brain that generate a conscious mind. A physical world that can generate emergent properties is a mysterious and magical world indeed.

What gets left out in such consideration is the capacity for personal control over one's biology, which is an important theme of this book. I contend that at the level of the individual person, mind itself—especially conscious mind—is a major force of natural selection that drives creation of mental capacity and character. To make that case, I will explain as simply as I can what neuroscientists know about how the brain works (chapter 2). Other books that discuss brain do so as if knowing how the brain works is an end in itself. I focus on the implications of such knowledge. Then I try to explain what consciousness is, what causes it, and its various states (chapter 3). Most importantly, in chapter 4, I challenge the position of many fellow neuroscientists who hold that consciousness is only an "observer" that cannot do anything, much less generate what we commonly call "free will."

The implications for daily living could not be more profound. Accepting one's biology and circumstance breeds helplessness and fatalism. So, it boils down to one's belief system. Either you are "captain of your own ship, master of your own fate," or you are shackled by the belief that change is not possible. Are we victims of biology and fate? This book will show both how the brain shapes its own destiny and how what you think and do shapes brain function.

CHAPTER 2

HOW BRAINS WORK

In the beginning, as a fetus, I did not have to find my mind. It was always there and it grew as I grew. In defense of such a statement, I should define what I think mind is. Mind is a biological system for information detection, processing, and decision making. Mind is also a brain state, arising from the brain's functional anatomy.

What I did have to find was my *conscious* sense of myself, a mind of which I eventually became aware. That took a few years.

I live out in the countryside. I often look out at an old, gnarled tree in my field about fifty yards from my patio. The tree was there when my wife, Doris, and I moved out to the country some forty years ago. I never paid much attention to that tree then. I guess it was young and looked like all the other trees. But today, it stands out—it has character, what with its drooping branches, gnarled galls, and sparseness of leaves. It reminds me of me, as I, too, am getting old but still hanging in there, galls and all. I, too, have a kind of unique character. So do you.

Our brains typically improve their internal structure over the years as a result of vast increases in the number and strength of functional connections, or synapses, between neurons. Most neurons look like microscopic trees, with branches that shoot out from their central bodies to make functional contact with the branches of adjacent neurons. Even in adults the number and strength of connective synapses grows with learning. This is where memories are stored. This is how you physically change your brain by what you think and do.

This chapter may tell you more about how brains work than you want to know. If that is a problem, you can gloss over it. If you do decide to browse casually through this chapter, make sure that you at least get the general idea of the "brain's currency" and circuit impulse patterns (CIPs). You will need this background to accept and appreciate the value of what is presented in the following chapters. Besides, you might learn some interesting things here that you somehow missed out on in school or college. All through my school and college experiences in life-science courses, the nervous system was covered at the end of a semester. As a result, the subject got short shrift because time to cover it ran out. My impression is that a similar problem exists in education today. You may know more about DNA than you do your own brain.

NEURAL NETWORKS: THE ARCHITECTURE OF COMPLEX SYSTEMS

Brain functions fall into the mathematical category of "complex systems." Such systems are envisioned as networks consisting of "nodes" that are widely distributed in space and time, and are highly interconnected and interactive. Recent imaging studies of adult humans reveal how different areas of the brain are hooked together (see figure 2.1). Harvard is currently engaged in a huge government-funded project to map the detailed anatomy of connections among various brain areas.

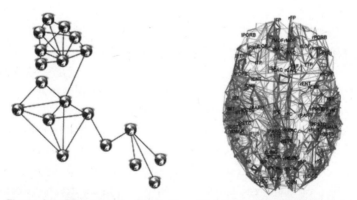

Figure 2.1. The network organization of brains. *Left:* schematic diagram of a brain network, showing pools of neurons as widely interconnected nodes. *Right:* an actual diagram of the connections of "nodes" (clusters of neurons) in the cerebral cortex of humans. (P. Hagmann et al., "Mapping the Structural Core of Human Cerebral Cortex," *PLOS Biology* 6, no. 7 [2008]: e159: 0001–0015.)

As you can see in the figure, not every area connects directly to all the others. The nodes in the left-hand drawing should be thought of as neural circuits, and thus the diagram depicts the brain as a network of interconnected networks. This network architecture has functional consequences. For example, if each cluster of neurons in the brain made direct connections with every other cluster, individual clusters would have little independence, and it would be hard to create a division of labor, which requires that individual clusters perform specialized functions.

The "wiring diagram" of the brain is implemented in the womb, resulting from the birth of billions of neurons that migrate to specified locations and make functional connections with certain other neurons. How do migrating neurons "know" what their targets should be? Our understanding here is fuzzy, but it seems clear that fetal neurons achieve individual chemical identity and that their unique chemical "signatures" serve as cues for directing migration. Migrating neurons that do not find "home" connections die, and they die by the billions in the fetus. Nobel-Prize winner Gerald Edelman calls this process "neuronal group selection" and "neural Darwinism."[1] Whatever you call it, this is the process by which brain is formed.

The way nodes are linked in the brain provides it with the anatomy to support complex-system-like properties that become manifest in important ways. First is the matter of *resilience*.[2] Perturbations of any one part of the network may be accommodated by adjustments elsewhere. I give some specific examples at the end of this chapter when discussing neural plasticity. When a system is highly resilient, it recovers rapidly from perturbations, and this is exactly what a conscious, thinking brain does. However, when a system loses resilience, it slows down near functional-state tipping points and becomes less able to change state. This is what happens when you fall asleep or wake up from sleep. It may take us all night to wake up, with the night being punctuated by aborted attempts to awaken (see the section called "Activated Sleep" in chapter 3).

Even physical damage may be partially compensated for by gradual rewiring of the connections, but of course this takes more time because a damaged system is less resilient. There are limits to what can be repaired in the brain. For example, if brain damage destroys your speech center as a young child, you may recover some speech capability, but not if the damage occurs as an adult. Young brains are more resilient than old ones.

Consider the property of *tipping point*. Perturbations in a highly interconnected system may build up to a tipping point that precipitates a domino effect throughout the network to create a systemic transition (as when we fall asleep or suddenly wake up). You might think of this like a contagion that spreads rapidly throughout the system.

Such arrangements can lead to sudden collapse of an ongoing state. Epilepsy is a good example. Here, a local damage to a few neurons can trigger changes that ripple throughout the system in a highly disruptive way.

The kinds of interactions among network nodes, or pools of neurons, determine the network's robustness. Interacting nodes may be mutually supportive or competitive, and that in turn determines the likelihood of collapse of a given mental state. Here again, epilepsy is a good example. Under most conditions, a focus of brain damage is held in check by inhibition from other nodes in the network. But any time the inhibition temporarily fails, extreme activity is unleashed, leading to functional collapse of the system that becomes manifest in a bodily seizure.

Modest changes occur in normal states. For example, when we are in normal sleep, the cerebral cortex is held in check by inhibitory influences arising out of the thalamus. But there is an arousal system in the brainstem that becomes active at the end of sleep, as when the alarm clock goes off. The arousal system removes inhibition of neurons in the thalamus, a major sensory processing center that connects to the outer mantle of the brain, the neocortex. "Cortex" simply means the outer surface of the brain, and all animals have one. But the human cortex has evolved a very sophisticated circuitry, and therefore it is called not just cortex, but *neo*cortex. The cortex is released into the state of consciousness. As in the case of epilepsy, a mental state has collapsed, but unlike epilepsy, normal collapse is more constrained and regulated. In both cases, the mental state changes upon reaching the tipping point of functional collapse. The difference between normal and abnormal collapse is that the brain evolved a capacity to tolerate certain kinds of state collapse, such as sleep and waking, because of biological necessity. Other state changes, like epilepsy, schizophrenia, clinical depression, and a host of other states are responses to disease.

TOPOGRAPHICAL MAPPING

A key aspect of neuronal architecture is that the body is mapped in the neocortex and its connecting pathways. Neuron assemblies exist as maps of all body parts for both sensation and control of movements. For example, the brain uses sense of touch, muscle sense, vision, and even hearing to construct a sensory image of its body and its interaction with the outside world. Similarly, there is another map for controlling body movements in specific ways. Historically, these maps have been thought of as a little person (a homunculus) residing in the neocortex.

Map formation means that sensations coming in from outside the body are registered in the part of the body map that is hardwired to the body part that was stimulated. Likewise, when the brain decides to move a particular body part, the neurons that control that body part are activated. For example, there is a part of the neocortex that maps touch and pressure sensations, so that one cluster of neurons responds to input from the toes, an adjacent cluster responds to input from the foot, another the calf, another the leg, and so on. In this way, the brain knows where information is coming from and in the process can reference input to its own body, which occurs in the context of a sense of self. In like manner, visual input goes through a distinctly mapped set of pathways to end up at the back of the neocortex. Sounds are routed to the cortex in the temple region.

The cortex also maps output to muscles, so that specific neurons activate the toes, the foot, leg, and so on. Thus, the brain has a sense of where in the body to route commands for action. More fundamentally, the brain uses its body maps as the core way in which the self is created and engaged with the world.

Neurons that are anatomically organized to function in a mapped way are called *topographical* maps. That is, each body part has cortical neurons that are dedicated to receiving sensation from it and delivering movement instructions to it. Even many lower animals, especially mammals, have well-defined sensory and motor body maps in their cortexes, but not all body parts are well represented, and the maps are most precise in humans. All mammals have well-defined topographical maps in their cerebral cortexes, though the size, shape, and relative location of the topographical maps differ by species. For example, in humans a relatively large expanse of cortex is devoted to the thumb, whereas in pigs a large amount is devoted to the snout. Similar principles apply also to the size of cortical tissue that is mapped to the body for commanding complex movements. Humans have many neurons devoted to controlling movement of fingers but relatively few neurons devoted to making back muscles move.

When we compare species, we have to look not only at topographical mapping but also at the total expanse of cortical tissue. The human cortex is only about 15 percent thicker than that of the macaque monkey, but it is at least ten times greater in area. We know from computers that there is more than a linear relationship between number of computer elements and computational capacity. At some point, increasing the number of computing elements gives rise to qualitative differences in capability. The same principle must hold also for the brain's computing elements, neurons. At some point, given enough properly functioning neurons, you reach critical mass for producing consciousness. In case you are not compelled by this "critical mass" and threshold idea, recall that the DNA of humans and chimpanzees is 98 percent identical and that the amino-acid sequences are 99.6 percent identical.[3] Amount of cortex is not all that matters. How the circuits within it are arranged may be even more important (see chapter 3).

In the process of mapping itself in relation to the environment, the brain creates the self-identity in which all other operations operate. Mapping provides a mechanism by which a brain learns about its bodily embedded self. Nerve impulses flowing in interacting maps constitute a representation of selfhood. My brain created a representation of me that it released into the world to act on its behalf. In chapter 4, I will develop this theme. Note that this self-identification begins in the fetus, where the mapped circuits are first formed. Functional deployment of the maps, sensory or motor, is performed consciously and unconsciously. How much conscious awareness occurs in the late-term fetus is not known, but monitoring of the fetus in higher animals shows that sleep and signs of dreaming occur, both of which are associated with the capacity for consciousness (see chapter 3).

Within certain limits, the brain can even reorganize its body maps if it has to. For example, studies in monkeys that had a nerve in the arm experimentally severed show that the part of the cortex that controls the originally innervated muscles rewired itself to take into account that those muscles no longer seemed to exist. Many neurons that formerly controlled those muscles became recruited to assist and expand innervation of other muscles in that limb.

Since body mapping of information is a fundamental property of brains, such mapping must have something to do with what we call mind. How this works is not known, but the leading speculation is that neuronal activity in brain maps must be recurrent. To illustrate, think of two pools of neurons, each receiving its own inputs yet interacting with the other in a mapped way (see figure 2.2). The communication is greatly enriched, perhaps to the point of making it accessible to consciousness, if the input in each map is relayed to the other map, which, in turn, sends it back to the first map. This process may occur in an oscillating way that would of course further magnify the effect.

Note that I suggested that recurrent activity among maps may make information *accessible* to consciousness. An alternative view is that such recurrent activity and oscillatory time may actually *be* consciousness itself.

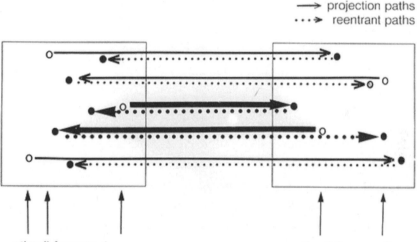

Figure 2.2. Schematic representation of selected neurons in two interacting topographical maps. Stimuli that feed into map 1 activate specific neurons (open circles) that project outputs (solid lines) to specific neurons in map 2. A similar process operates with stimuli that feed into map 2 (black circles). In both maps, the neurons that receive input from the other map deliver feedback ("reentry") into the original map (dotted lines). This arrangement could create oscillatory processes. Also, if stimuli are repeated, the synapses involved in that particular pathway can be strengthened (heavy lines), creating a basis for memory of the stimuli. (Figure by the author, based on data from *Bright Air* by Gerald M. Edelman [New York: Basic Books, 1993].)

Such "locally mapped" responses may be extended by interactivity with other mapped circuits, leading to a more "global" mapping that allows concept categorization and holistic processing. Expression of globally mapped concepts yields language, behavior, and consciousness. Some scholars call this global mapping the *global mental workspace*.

Nobel laureate Gerald Edelman proposes that higher cognitive functions involve a global mapping wherein maps interact with other maps. Various body maps do have reciprocal connections, and collectively their interactions could be expected to produce the superadditive properties necessary for emergent consciousness. The topographical maps in the neocortex also interact with brain maps in subcortical and brainstem structures, thus providing a gateway for engagement with all body functions. Subcortical maps don't necessarily map the body, but they send nerve fibers from one cluster to a target cluster in specifically arranged ways. Whatever the case, the point is that these mappings create the biological basis for distinguishing self from non-self.

The practical point of mapping is not just the creation of wiring diagrams as a developmental process in the fetus and children. Through our experiences, the choices we make, and the thoughts we indulge, we are constantly modifying our brain maps.

In this scheme, the operational units of the nervous system are not single neurons but rather groups of strongly interconnected neurons—in other words, circuits. Membership in any given functional ensemble is determined by synaptic connection strengths, which can change in response to learning. Connection strength depends on how many synaptic contacts there are, how robustly they can manufacture and release neurotransmitter, and the number of postsynaptic receptor molecules and their sensitivity to transmitter. All these processes are altered by experience. As distinct from the large neuronal groups that make up a map, these smaller groups are the units of "selection." That is, specific patterns of sensory input "select" a subset of neurons and their interconnections to represent the categorization of that input. The dynamic competitive process of neuronal group selection provides the brain with a way to change its circuits. This is also how brains learn and remember. You should think of the brain as a computer that programs itself.

Certain brain circuits are set aside for certain kinds of innervation. In the neocortex, as mentioned, there are areas dedicated to hearing, speech, and vision. The vision area is subdivided; that is, there are cells that recognize faces, areas that recognize where a visual object is in space, and another area that determines what the object is. We could go into mind-numbing detail, but the point is that there is a place for everything in the brain and everything has its place. But the places share what they experience and know.

Many of these mappings are learned through life experience. A professional violinist, for example, will have a larger than normal map in the neocortex for the fingers. If a nerve from a certain body part is cut, that part is no longer represented in the cortex. The cortical neurons that formerly innervated that body part get functionally recruited into nearby circuits that innervate nearby body parts. In a glib sort of way, this is the explanation for why blind people have compensatory enhancement of other senses.

These neuronal groups exhibit redundancy in that different circuits can have similar functions, or a given neuronal group can participate in multiple functions, depending on the current state of its dynamically changeable connections within and outside of the map. Here, I need to expand on the ideas illustrated in figures 2.1 and 2.2. Though these diagrams show hardwired circuits, they do not indicate that one or more of these circuits can dynamically change (in the sense that all or part of a given circuit can be transiently recruited to participate in the function of other circuits from which they receive input). This kind of grouping might be the main reason the human brain is so powerful.

Mapping phenomena seem to have a straightforward relationship to sensory and motor processing. But for certain "higher" functions, such as emotions and memory, it is not clear what role is played by mapping. However, I should mention that the hippocampus, which helps form memory, also maps the environmental space in which learning occurs. It seems likely that time is mapped in dynamic patterns of rhythmic activity. Maybe higher functions are mapped in terms of their information category rather than in terms of specific body anatomy. Emotions are clearly controlled by a complex, multiply interconnected set of structures, known as the limbic system, located underneath the neocortex. The limbic system dominates the brain of lower animals, but it acquired direct connections with the neocortex in the evolution of primates. Even though limbic structures do not map the body, as such, many parts of this system show point-to-point mapping. Such nonbody maps are still important because they constrain what connects to what. Nature does not allow neurons to project willy-nilly just anywhere.

THE BRAIN'S CURRENCY

Issues relating conscious mind to the material world were clarified by classical philosophers, David Hume, John Locke, and Immanuel Kant.[4] Kant made forceful arguments that all human knowledge begins with experience, but not all knowledge is derived from experience. Perhaps Kant's most famous insight was that human reason cannot answer all questions about the true nature of reality (see chapter 5).

It is clear that sensory experience alone is inadequate (this limitation probably applies also to the instruments and tools of science). Thus, reason based on sensation is inadequate for discerning reality. Three hundred years after Kant, we can now confirm from the neuroscience of our time that brains create abstracted representations of the world. Because reality is abstracted, you cannot detect all there is (for example, you don't see gamma rays, and you don't hear all of the vocalizations of various animal species). As I discuss in the last chapter, even our scientific instruments do not detect everything we know must exist.

Looking at my old tree in the field causes me to ask, "What is it I see—that is, my brain sees—as my eyes fall nostalgically on its galls and withering branches?" Visual images that are detected by the eye are not sent into the brain like some movie screen that is painted on the visual cortex. As I explain below, the visual image at the retina of the eye is broken up into little pieces of image; each piece is actually a small line segment that is conveyed into the brain by its own small set of neurons. Then the brain puts all the little pieces together again to reconstruct a mental representation of what the eye saw.

Actually, my brain does not see the tree, as such. It sees an abstract representation of the itty-bitty pieces of it. The little-piece representations exist in the form of patterns of electrical pulses, called nerve impulses, flowing as voltage pulses in dynamically adjusting circuits. The tree is first represented to me by the wavelengths of green, brown, and shades of gray light that bounce off the tree. The retina of my eye then absorbs these wavelengths and in turn generates impulses that get conveyed into my brain, ultimately reaching regions at the back of my head that are collectively called the *visual cortex*. In this example, the visual cortex generates the mental images of my mind.

The point is that these impulses of electricity are the carriers of the information about the tree.[5] Impulse voltage pulses propagate along neuron processes from one neuron to another or to target cells like those in muscles and glands. In the case of my old, gnarled tree, my brain cannot see a tree. It sees an abstract representation of small segments of the tree that together, when pooled over thousands of neurons along that visual pathway, fabricate an image. The tree exists, of course, but in my mind it is a figment of what I imagine the tree to be like on the basis of how it is represented in my brain. Actually, my eyes only detect a portion of the light that is reflected from the tree. I don't see, for example, the tree's infrared reflection.

What is so amazing is that each cell in my retina and along the visual pathways is carrying only a tiny bit of this representation. In the visual processing area of the cerebral cortex, what is represented by any one cell is simply bars and edges, and very small ones at that.

We know this is the way things work because of two 1981 Nobel-Prize winners, David Hubel and Torsten Weisel, who put microelectrodes into the visual cortex of monkeys and recorded how each neuron responded to visual stimuli, which they controlled as small black bars on a white background. A given neuron in the visual cortex responded only to the bars in a certain spatial location and orientation. For example, response depended on the horizontal angle of the bar. If the visual stimulus were rotated only a few degrees, the neuron would stop discharging impulses.

Impulses are voltage spikes on the order of one millisecond long that propagate along neuron processes, from one neuron to another, or to target cells like those in muscles and glands. The current that generates these voltage spikes is carried by two atoms, sodium and potassium, that have lost an electron to water and thus are ionized. This, by the way, is a basic reason why living things need a solvent like water—to enable ionization.

The patterns of photons from the tree that strike my retina are represented in my brain by such impulses. Yet this basic fact is sometimes misunderstood by people who should know better. I am reminded of the recent book *Physics in Mind*, in which physicist Werner Lowenstein argued for the relevance of quantum mechanics to the human mind based on how eye pigment handles photons. But the brain does not run on photons. Brain runs on sodium and potassium ions, with an assist from calcium ions, as I documented in my recent book *Atoms of Mind* (Springer). There are no photons inside nerve cells. It's dark in there.

While the retina processes photons, the brain processes the information content provided by the atoms that create nerve impulses. Actually, the atoms involved in impulses are ions; that is, they are positively charged because they have released an electron into the water solvent in which they reside. Brains run on liquid-state electronics, not the solid-state electronics of computers.

The Nobel Prize was awarded in 1963 to Alan Hodgkin and Andrew Huxley for showing that impulses are created by flow of sodium ions into a cell and a subsequent flow of potassium ions out of the cell. Both flows are completed in about a thousandth of a second, and thus the brain's information exists as pulses of this ionic electricity. As an aside, others showed later that positive calcium ions are important to information flow in that when they flow into nerve cell terminals they stimulate the release of neurotransmitter chemicals that modulate impulse propagation from one nerve cell to another.

We don't need to go into the details of how these phenomena were proved. But it is instructive at this point to understand a few principles of electricity, bioelectricity in particular. The electrical signaling in the power lines supplying your house is carried by electrons flowing in conductive wire. In your computer, electrons hop around and through semiconducting solids. Such solid-state electronics are quite different from electrical signaling in living systems, where the electricity is liquid state.

There is electron flow in the brain, but it is involved in energy production, which is similar to that found even in primitive organisms. In mitochondria, the "energy factories" of all cells, for example, electrons hop around in a solvent medium (water) from temporary anchor points (metals, often iron, bound inside certain proteins) that help guide electron flow in an orderly manner. This typically occurs within each mitochondrion, and the whipping around of electrons generates kinetic energy for the organism. Mitochondria are ubiquitous in all higher organisms and are especially prominent in neurons, which demand a great deal of energy.[6]

So why didn't electrons get used for signaling in brains? I think it is because this kind of signaling is too crude, too diffuse. It is akin to lightning striking a lake: the electrons just splatter everywhere or are only loosely constrained. Natural-selection forces reward better nervous systems, and mammals, primates, and humans are here to prove it.

In our computers, flow of electrons serves useful purposes by the way we constrain electron flow in the design and architecture of computer chips. A main reason higher animals are "higher" is that they, too, have a way to constrain and direct electrical flow rather than allowing it to splatter everywhere. Animal and human nervous systems have to use a different method because we are 97 percent water-based, which would, if that was all there is to it, just allow the electricity to splatter diffusely.

Nervous systems deal with this problem in two ways: (1) the signal carrier is atoms that ionize in water, and (2) electrical flow is constrained by confining the direction of impulse propagation throughout neural networks. This realization is sometimes difficult for physical scientists to grasp, as in the proposed photon mechanism mentioned above. I remember getting into an argument with a physicist after a guest lecture in Hawaii. He accepted the fact of positive ions as the carriers of information in the nervous system. But he insisted that the free electrons and the voltage fields they generated as they flowed through the electrical resistance of tissue were also participating. (Actually the voltage fields do influence the activity of neurons that generate them, but the fields come from ion flow, not electron flow). He never really bought into the several facts that undercut his position:

- The voltage fields are primarily created by ionic current, not electrons.
- Free electrons in tissue could in theory flow most anywhere, much like lightning in a lake. Optimal signaling and information processing require electrical flow to be orderly, constrained, and directed.
- Free electrons in tissue don't stay free for long; they are instantly grabbed by any molecule that is short of electrons. Typically these are intracellular proteins that actually collect a surplus of electrons and are the main reason that all resting cells sustain an internal electrical negativity.

It is true that the flows of ionic current in extracellular fluid generate voltage fields that can impose bias on the generation of impulses. But this is a bias on genesis and propagation of information, not the information itself, which resides in impulses.

Impulses can be triggered by stimulation or by certain chemicals. A neuron at rest, like all cells, is an electrical battery, polarized, with the inside of the cell electrically negative relative to the outside. The "battery" of neurons, however, can be discharged (depolarized) when membranes become more permeable to sodium ions, yielding an impulse. Electrical fields and certain chemicals can alter membrane permeability.

Impulses carry information into the brain from all sense organs. Impulses carry information inside the brain's millions of circuits during the processing needed to store memories and generate intentions, choices, and decisions. Scientists can't yet read the "neural codes," but it is clear that information is represented in neurons in terms of onset and offset of impulse generation, number of impulses, and intervals between impulses. Within a given circuit of connected neurons, we can think of information as being represented in terms of circuit impulse patterns (CIPs). At any given instant, each neuron within a circuit may or may not be generating impulses. It is possible that simultaneous activity across all neurons in a circuit at a given point in time might have some sort of combinatorial or collective code. Scientists haven't given much thought to that possibility, in spite of the obvious fact that activity across an entire circuit contains much more information than impulse discharge in any one of the circuit's parts.

If there is no impulse flow, there is no ongoing thought. As with computers, if you turn off the electric current, the computer becomes nonfunctional. In humans, the point has been made experimentally: for example, one can inject an anesthetic into a carotid artery and disrupt all impulse traffic—and the corresponding thoughts—in the area of the cerebral cortex supplied by that artery.

There is *latent* thought however. In the above example, those anesthetized cortical circuits still have a capacity for thoughts stored as memories in their synapses and connection pathways. In other words, they have not lost their capacity for thinking. However, thought itself is not *expressed*, because the anesthetic disrupted impulse discharge. The computer analogy applies here also. Information in a computer resides in stored hard-disk memory, but it can be expressed only when the computer is turned on.

The relationship of anatomy and biochemistry to the human mind, compared with nerve impulses, can be likened to the relationship of potential energy to kinetic energy. One brain state represents the capacity and storage form for the mind while the other reflects mind in action. Thoughts, for example, can reside in latent form in the microanatomy of neural circuits and their associated synaptic biochemistry or they may be expressed and ongoing in the form of the circuit impulse patterns (CIPs). When a person is awake, the CIPs constitute an actively deployed, "on-line" mind that interacts with the world. This is also the mind that programs what goes into storage (memory) for later deployment as the "up-dated" on-line mind. Remember this CIP idea. It crops up again and again throughout the book.

Once triggered into being, impulses spread throughout a neuron like a burning fuse and extend into all that cell's terminals, which provide a way for one neuron to communicate with many others at roughly the same time. Actually, because a given neuron has numerous terminals, its impulses may reach hundreds of other neurons; many even spread into other network nodes that are not close neighbors.

Brains have two basic kinds of cells: (1) neurons, and (2) cell types, called *glia*, that support the neurons. As far as we know, the primary cause of brain function and mind comes from neurons. However, at least one type of glial cell discharges long-duration calcium-based voltages on the order of seconds that could bias the excitability of neurons in the immediate vicinity.[7] Glia, estimated to exceed neurons in number by about ten-fold, surely have other intimate relations with the neurons they surround, but much remains to be learned about what glia do.

Neurons are organized into circuits, formed either under genetic control or under influence of environmental stimuli and learning. Some of these circuits are in constant states of flux, turning off and on, becoming more or less active, and changing the activity of their constituent neurons, impulse firing patterns, and routing pathways. Circuits interface with other circuits, as for example one circuit being embedded within other circuits.

The most important set of circuits in higher animals is the "cortical column," which is a set of circuits in a perpendicular orientation to the neocortical surface and surrounded by other similar sets of circuits. Microelectrode studies have identified the principle components (see figure 2.3).[8] The process involved stimulating one neuron a known distance from the cortical surface and noting which neurons responded to the stimulus and their locations. Under normal circumstances, activity may be going on in parallel in all the column circuits, and the circuits may interact with each other via feedback (see pathways D, E, and F in the figure). This constellation of multiple processes underlies what is called thinking. In other words, thoughts are CIPs.

The information at any given instant is contained in all the impulses in the set of circuits, not just what is happening in any one neuron or even in one isolated circuit. The impulse coding may be combinatorial, that is, contained as the impulse patterns in all the neurons of the circuit at a given moment. The information content of a given circuit may change as it receives particular input from other circuits in the brain and the spinal cord: some neurons may drop out of the circuit, while others may be recruited. Many neurons are shared by multiple circuits, as for example, the three neurons in circuit A are also part of circuits B and C.

Many neural circuits are hardwired and perform a predictable behavior when activated. An obvious example is the knee-jerk reflex. Most people have had a physician tap on a knee tendon to observe the magnitude of knee jerk. Another example is the swallowing reflex, which automatically lets you swallow food when it arrives at the back of the throat. An associated reflex causes tissue to fold over the larynx so you don't gag. Other neuronal circuits are malleable and can be constructed on the fly, so to speak, depending on the needs of the moment. Such circuits enable learning and memory of all the many experiences in life for which the brain was not born with a dedicated hardwired circuit. Thank goodness. Your brain would have to be larger than a barn if every experience demanded its own hardwired circuit.

Scientists have traditionally studied neurons by using small electrodes to record voltage changes, both the relatively slow, graded voltage changes in junction points (synapses) and the pulsatile nerve impulses. If a relatively large electrode, about a third the diameter of a dime, for example, is placed on the scalp, it will sense the voltage changes of thousands of neurons in the region of brain closest to the scalp. This is what we call the electroencephalogram or EEG. There is a magnetic field counterpart, recorded as a magnetoencephalogram. The EEG thus can monitor thinking states of large populations of neurons in the outer neocortex mantle of brain, which is the part of the brain that provides most of the electrical signal at the scalp, does the most sophisticated thinking, and gives rise to conscious awareness.

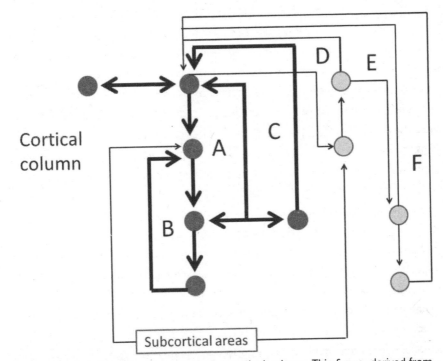

Figure 2.3. Simplified neural circuits in the cortical column. This figure, derived from microelectrode stimulus-and-response recordings of single neurons in known layers of the cortex, is central to all that follows in this book. The key ideas are that neurons (represented here as circles with projecting lines) are organized into circuits in which information is propagated in the form of patterns of nerve impulses. Neurons have cell bodies (circles in drawing) and membranous processes (represented here as lines) that project to other neurons. All neurons indicated here are excitatory (these circuits do interface with some inhibitory neurons, which are not shown for simplicity). The heavy lines indicate primary neurons in the cortical column and the lighter lines and circles indicate connections with neurons in other cortical columns. Impulses typically propagate outward from cell bodies to act on target neurons. Such circuitry illustrates parallel distributed processing.

Circuits do not operate in isolation. Some six interconnected circuits are shown here (A–F). They receive inputs from and send outputs to other circuits at various points.

(Figure by the author, based on data from R. J. Douglas and K. A. Martin, "Neuronal Circuits of the Neocortex," *Annual Review of Neuroscience* 27 [2004]: 419–51.)

Heightened consciousness is the hallmark of humans, and the human brain differs most conspicuously from other animals in the development of the brain's outer surface, the neocortex. Indeed, the reason it is called a neocortex, or a new kind of cortex, instead of just cortex, is that it is regarded as the evolutionarily most recent cortex. Circuit organization is also different in the neocortex.

If an electrode is very small, like a micro version of the tip of a sharpened pencil, and is thrust directly into the brain, it will detect the net electrical activity of a dozen or more neurons in the immediate vicinity. Scientists call this "multiple-unit activity" because the activity comes from multiple neurons. If you insert into the brain an electrode that is less than the diameter of a human hair, the electrode may detect only the voltage of the nearest neighbor, a single neuron. If you impale a single neuron with such a microelectrode, the electrode will not only detect the impulses from that neuron but will also reveal slow-graded excitatory and inhibitory modulations of the voltage (postsynaptic potentials) across that neuron's cell membrane as it responds to input at its synapses. Like a battery, the neuron is electrically polarized. Excitatory input accumulates to depolarize it and generate impulses. Inhibitory input does the opposite. If, in addition, you use a glass capillary microelectrode and suck a small patch of membrane into the tip, you can even detect ionic currents flowing through ion channels as they open and close in response to input to the neuron.

Currently, President Obama is trying to launch a multibillion dollar "big science" program to map neural impulses simultaneously from up to a million neurons.[9] This is a boondoggle in the making. Such a project is naïve and overly ambitious. Information coding in neurons is not like the coding of genes. Yes, each defined circuit contains a code, probably a combinatorial code of impulse patterns in all the neurons of the circuit. But there are many millions of circuits and thus many millions of codes. Moreover, the codes change with changes in brain state, and in recalled memory and thought.

Such research may take us further away from understanding the larger matter of thinking. In that sense, such methods may only teach us more and more about less and less. Science makes it easy to get lost in tedious detail while missing the big picture.

Then there is the anticipated huge cost. The drain of federal funding will divert much-needed support for more traditional neuroscience research and, of course, other kinds of research. The proponents' argument that this program would be comparable to the human genome project is specious at best.

The proposal rightly has a focus on the most important aspect of human mind: neuronal impulse trains. But what is needed instead of a shotgun fishing expedition are systematic small-scale studies involving recording and combinatorial analysis of spike trains from all neurons within defined circuits during a variety of cognitive conditions.

Brain circuits are arranged not only in series but also in parallel. Multiple operations can go on simultaneously in multiple parallel circuits. This kind of processing is especially prominent in the neocortex. The basic structure of the cortex of primates is built upon multiple cortical columns, each containing defined layers of neurons arranged perpendicular to the cortical surface. Within each column, the neuron cells' types have been identified, along with their connections with the other cells in the column.

Microscopic examination of the neocortex shows that all parts of it have a similar miniature-network organization of their columns.[10] The neocortex of humans has six layers, and inputs terminate in specific layers while output projections arise in other specific layers. The neocortex has an outer surface layer of fibers. Beneath that layer are arranged neurons that are primary targets of input fibers. Some column neurons send output projections to the spinal cord or other parts of the neocortex, or to other parts of the brain, or to interneurons that modulate activity of the other neocortical neurons. What gives the neocortex its localization of specific functions is the source of the mapped inputs.

Much of this was determined by microelectrode studies in cats and monkeys; the details have not been confirmed in humans. Though the human neocortex shows similar anatomical layering and cell types, there may be differences from other primates in the connections among cortical neurons.

The rich interconnections of various neocortical areas (see figure 2.1) provide a way for the whole complex to operate as one unit, despite all the diversity and specificity of various CIPs contained in local column microcircuits. I think that a unitary mode of operation across multiple areas of the neocortex is what produces consciousness. Think of consciousness as a system property of the brain; that is, the simultaneous engagement of the whole neocortex allows it the opportunity to have a sense of self that is aware of what any particular part of neocortex is doing (seeing, hearing, deciding, planning, and so forth).

One consequence of the cortical-column organization is that excitatory input arriving in a given column can ripple through adjacent columns like ripples that move through a pool of water after a rock is dropped in. Such rippling has been observed in so-called spreading-depression experiments in which applying potassium chloride to any spot on the cortex causes neural-depression activity at that spot that then spreads to shut down the whole cortex.

At the same time that input spreads laterally, column by column, there is an accompanying further communication with many other parts of the neocortex by way of the fiber tracts within and between the brain's two hemispheres (recall the second diagram in figure 2.1). This is parallel processing on steroids. If allowed to go out of control, an excitatory ripple effect can result in an epileptic seizure that captures the function of the entire brain.

This elemental circuit design includes recurrent excitatory and inhibitory connections within and between layers. Most of the excitatory drive is generated by local recurrent connections within the cortical layers, and the sensory inputs from the outside world are relatively sparse. The usefulness of this design is that weak sensory inputs are amplified by local positive feedback. The risk of such organization is runaway excitation, and the problem emerges when pathology removes the normal inhibitory influences that hold the cortical circuitry in check.[11] Excitation that is similar, but less runaway than what occurs in epilepsy, occurs in many neurological diseases, such as Parkinson's disease, Tourette's syndrome, or even compulsive disorders.

Clearly, such organization suggests that cortical columns are mutual regulators. Clusters of adjacent columns can stabilize and become basins of oscillating attraction, and the output to remote regions of the cortex can facilitate synchronization with distant basins of attraction.

Control in such a system is collective and cooperative. Although you might speculate that such executive control could create consciousness as a top-down or even nonphysical "ghost in the machine" process, the facts argue for consciousness emerging from the spontaneous action of mutually regulating circuits that in one state operate as conscious mind.

The amount of neocortex in humans is relatively much larger than in other primates. Maybe the unique human thinking abilities are only possible because of a certain "critical mass" of those modular-circuit columns.

The idea of nerve impulse patterns as information representation is key to developing the proper understanding of mind. But detecting and quantifying impulses in a single neuron is not sufficient. Thought is represented by the spike trains from all the neurons in a given circuit at roughly the same time. Thus, whatever code neurons use for thinking, it must capture what is going on in defined circuits as a collective; that is, all impulse activity in those circuits.

Nerve impulses are the only things that propagate throughout the brain's circuitry. As such, they constitute the messages that are being routed throughout the brain. "Information processing" occurs mostly via neurotransmitters (see below) that operate in the junctions between adjacent neurons, but the transmitters do not propagate their chemical signal over more than a few microns. Postsynaptic receptors and biochemical amplifier systems are also confined to synapses. Postsynaptic voltage changes do propagate for a few microns of space, but they do not move the many millimeters, even meters, that can be accomplished by impulses as they self-generate along neuronal processes that lead to other neurons.

In other words, nerve impulse patterns are the currency of thought, presumably at all levels of mind. Both unconscious and conscious functions include CIPs ranging from the fixed circuitry of those in the spinal cord and brainstem to dynamic assemblies of neocortex neurons whose functional connections come and go in the course of neural processing. No doubt, the richness of combinatorial coding in dynamic assemblies would be greatest in conscious mind.

The brain's unconscious mind includes certain operations that regulate emotions and their influence on brainstem neuroendocrine controls as well as on the neurons regulating such visceral functions as heart rate, blood pressure, and digestive functions. Unconscious operations also produce a wide range of movements that have become so well-learned by neural circuitry that the controls are automatic, much like your home-heating system, and can be performed in robot-like fashion without consciousness. These may support normal required functions or produce diseases like hypertension, palpitations, or ulcers if we are in chronic mental distress. But in all unconscious operations, it seems reasonable to assume that the representation for information and operating on it to generate responses or actions occurs in the form of combinatorially coded CIPs.

Neurons propagate their spike impulses through circuits such as those illustrated in figure 2.2. As impulses reach the synaptic junctions between neurons, they generate voltage fields in the synapses, causing chemicals (neurotransmitters) to be released to modify information flow. The circuits are embedded in the extracellular voltage fields that they generate, although many regions of the circuitry may be electrically insulated by surrounding and nurturing glia cells. In any case, impulse currents can contribute to an external voltage field that in turn affects the excitability of neurons within that field.

How do all of these impulses, voltage fields, chemical releases into synapses, and dynamically changing circuitry give rise to thoughts? They don't give rise to thoughts. Rather, they *are* the process of *thinking*. Thinking is equivalent to the CIPs. Why CIPs instead of what is happening in the synapses? First, thinking is a dynamic process, and its "messages" are propagated and distributed in real time in the form of CIPs. What happens in the synapses becomes expressed in the CIPs. The chemical and microanatomical changes that occur in synapses represent the memory storage and processing of thought. The expression of thought occurs as CIPs.

In any given neuron, the impulse patterns take the form of rate of firing and rate change in firing, onset and offset of firing, and sequential order of intervals. You could think of CIPs as instruction sets for performing such brain operations as detection, integration, decision-making, and commands (see figure 2.4).

Research in the last several years is exploding with evidence for the above view. For example, I realized in the early 1980s that nerve impulses did not occur randomly, that they contain a code, not only in the rate of firing, but also in their interval patterns. At that time a few other scientists had arrived at the same conclusion. More recently, I realized that oscillation and synchronicity of the more global electrical voltage fields were important (more about that later in the book). These new insights are not mine alone. In the last ten years, numerous researchers have been providing experimental data to support both ideas. Now the time is right to make these thoughts explicit and simply explain how they are supported by experiments. These ideas about CIPs and oscillation of circuit activity are central to understanding brain function in general and consciousness in particular.

Neurons can only cause one of two things at the cells they contact: either excitation or inhibition. If the target cell is a skeletal muscle, neurons can only excite it, causing it to contract. If the target cell is a smooth muscle, as in the gut, or another neuron, a neuron may excite or inhibit. The effect is dictated by the kind of chemical that neurons release at their terminals as impulses arrive in the terminal. These chemicals, called neurotransmitters, diffuse across the synaptic gap and attach to receptor molecules on the target neuron, and, depending

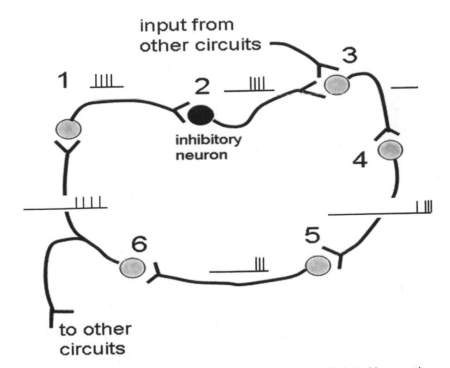

Figure 2.4. Illustration of the idea of circuit impulse patterns (CIPs). In this example of a small circuit, each neuron (circles) generates a certain temporal pattern of impulses that affects what happens in the target neuron (shown as spike patterns for each neuron). Note the time delay at each synapse and the pattern change in each neuron. There may be more than one entry or exit point from the circuit. For example, impulses from the inhibitory neuron (number 2) shuts down activity in neuron number 3, which nonetheless may reactivate when the inhibition wears off or when excitation comes from another circuit with which it interfaces. Collectively, all the neurons in the circuit constitute a CIP for a given period. When embedded within a network of interfacing circuits, such a CIP may become part of a more global set of CIPs. Such CIPs are regarded as a representation of specific mental states. The meaning of this representation may lie in the combination of impulses in all the circuit members, not just any one member neuron, contained in the form of some kind of combinatorial code.

on the chemical, change the flow of ions across the target's membrane that will either excite or inhibit it.

The brain has over a hundred neurotransmitter systems. You probably have heard of some of these transmitters, such as the "feel good" transmitters, dopamine and endogenous opioids. I say "systems" because for each transmitter there must be a series of chemical mechanisms to make the transmitter from precursor molecules, store it in neuronal terminals, and move it to the terminal membrane where it can be "spit out" into the synaptic gap in response to the voltage changes created as nerve impulses arrive in the terminal. The transmitter then must have a molecule that recognizes and binds it on the postsynaptic neuron. The binding must be reversible and relatively short lasting so that the target neuron can become responsive to other nerve impulses. Thus, there must be a chemical in the area that can break down the transmitter so that its action can be terminated. Finally, there is often a chemical mechanism to transport the transmitter or its components back into the presynaptic neuron to conserve the raw materials for making the transmitter available again.

To understand why patterns of impulses matter, you need to know that the amount of neurotransmitter released is proportional to the number of impulses arriving in a short time; that is, the more impulses that arrive in the neuron terminal, the more neurotransmitter will be released. Similarly, the more neurotransmitter that is present in a synapse, the more likely it is to change the activity of the target neuron. Target neurons, like all neurons, generate impulses when their membrane becomes destabilized to allow sodium ions to rush in. Binding of neurotransmitter produces that membrane destabilization. There is a threshold of voltage change that must be reached before the postsynaptic neuron discharges impulses. A little transmitter causes a little voltage change, which will decay unless another batch of transmitter is released soon. If several batches of transmitter are released at about the same time, the target neuron membrane may become sufficiently destabilized to reach the threshold for generating its own impulses. In other words, postsynaptic membrane changes summate (or accumulate) over time, and the degree of summation depends on the timing of impulses arriving from the presynaptic neuron. There is also summation over the number of synaptic junctions on a given neuron that are active at any one time.

Some synaptic junctions operate in a similar time- and space-summating way, except that the transmitters involved create an opposite effect on membrane stability. This inhibits impulse traffic in a network. The membrane becomes more resistant to destabilization, making it harder for the postsynaptic neuron to generate impulses—thus, it is inhibited.

Recall that I said earlier that CIPs constitute instruction sets. As in figure 2.4, the information content of the circuit is captured in the firing patterns of every neuron in the circuit. These, of course, change from moment to moment as the circuit activity is modulated by inputs from other circuits. How "instruction" is delivered to other circuits or muscles can be unbelievably complicated or sometimes quite simple. In the case of circuits in the part of the motor cortex that generates commands to skeletal muscles to contract, the instruction consists of the circuit delivering enough impulses in a short enough period to activate contraction in the target muscles. For complicated "instructions" in neuronal circuits, notice in figure 2.4 that I show one of the neurons (number 6, though there could be more) sending an unspecified output to another circuit. The impulses that neuron number 6 sends elsewhere are of course generated by what happened recently in neurons 1 through 5. The impulses that propagate from number 6 will either occur often enough in a short enough time to excite a target in the distant circuit, or will bias its target to be more or less responsive to other inputs, or will decay and fail to have an effect.

Neurons come with a wide variety of firing patterns. Even in a single brain area, intrinsic properties vary widely. In the hippocampus, which processes memory, for example, one cell type produces a short train of spikes that habituates after stimulation with a short depolarizing pulse but a single spike in response to a super-threshold pulse of current. Another type fires bursts of pulses in response to long and strong current pulses, but only a single spike in response to weak stimulation. Another type also fires bursts in response to weak but long stimulus pulses. Another type is similar, but its bursts are very stereotyped. Yet another type fires rhythmic bursts of spikes spontaneously without stimulation. While we don't know the functional message of such patterns, intuitively we believe that the various patterns do have distinct effects on their targets. For example, in an excitatory circuit, a neuron that delivers a sustained train of impulses to its targets is likely to generate a sustained response from the target. A neuron that fires in bursts might produce "on" and "off" responses in the target corresponding to when the target receives a burst and when it does not.

As I will elaborate later, the CIPs most relevant to conscious thinking come from neurons in the neocortex, and the circuits of these neurons are closely packed like small columns oriented more or less perpendicular to the brain surface. The electrical currents resulting from impulses in each column's circuitry readily summate as a collective consequence of the impulse activity in all members of the circuit. Often, this collective combinatorial effect drives frequency-specific oscillations of the whole circuit and neighboring column circuits.

THE NATURE OF THOUGHT

Thoughts originate from what we sense in the environment or from memories of such sensory experiences. They can also be creatively generated by mental operations on sensory input or memories.

What we sense or remember is

- an IMPLEMENTATION (in brain circuitry) of
- a REPRESENTATION (as nerve impulse patterns)
- of an ABSTRACTION
- of the actual REALITY (physical stimuli).

What we think and believe is based on

- how that sensory or memory representation is processed, remembered, and altered by feedback from our actions.

Every thought, conscious or otherwise, has a CIP representation. Is that the same as saying that the CIP *is* the thought? Why not? A thought has to be expressed in some way to be useful. If the thought can be visualized, then it becomes expressed when the visual cortex creates the image of the thought in the "mind's eye." If the thought can be described verbally, it becomes expressed when the CIPs include the language systems in the left hemisphere.

To think, the brain has to tag, match, or compare CIP representations with real-world phenomena. How? If the CIPs represent ongoing sensory stimuli, the matching is automatic; since the CIPs were generated by the stimuli, they therefore are an automatic representation of those stimuli. In the case of memory or internally generated information, those sources of input to the virtual working-memory scratch pad have their own CIPs, and these have to be matched and integrated, in short successive steps, with the CIPs that are being generated elsewhere.

The form such matching entails is not known, but it almost certainly relates to the fact that many circuits overlap physically. That is, one or more neurons in one circuit can be shared with other circuits. The impulses from one circuit mix with the impulses coming from the other circuit. Thus, in overlapping circuits information may be tagged by increased total activity. Or maybe the "tag" is a localized increase in activity or a shift in CIP time relationships.

Overlapping provides a way for one set of CIPs to be shared with the CIPs elsewhere. Moreover, each overlapping circuit has access to CIPs that are about to be generated in the immediate future and a back reference to the CIPs that are coming in from senses or that emanate from memory.

Such sharing of information allows each set of CIPs to "know" something of what the other set is doing (i.e., "thinking"). More importantly, the overlap allows exchange; that is, each set can help program the other, even though each can maintain a degree of autonomy.

How can this conjoint information be "read"? If the spike trains in one set of CIPs merge simultaneously with the trains from the other CIP, then the merged spike train would seem to be garbled, unless, of course, neurons merge the impulse patterns in some sequenced way that preserves where each CIP came from. The "time chopping" of oscillation could provide a way to let spike influence from one CIP enter into another while still being kept distinct. To keep from scrambling the messages when chunks of spike trains are mixed, specific interval clusters of spikes would need to remain intact, which would reflect a "byte" mode of information processing. Although the idea of spike-interval coding has long since been abandoned by most neuroscientists, the evidence for it has not been refuted.

Over thirty years ago, Cliff Sherry and I presented evidence for byte processing. We documented that neural circuits carry information in the form of spike clusters of serially dependent intervals.[12] We did this by computing the relative duration of adjacent intervals. Each interval was scored as being shorter (-), longer (+), or approximately the same (O) than the immediately following ones. Calculating the incidence of specific interval patterns revealed the occurrence of specific clusters at far greater levels than could occur by chance (for example, a nonchance cluster might be +++-, or 00-+, etc.). Mixing such input from two or more spike trains could produce an output that preserves the distinct packets (bytes) of information in ways that could be read and differentiated by other circuits to which it is projected.

This model is somewhat analogous to the genetic code. While most of the DNA does not directly code for making proteins, there are many isolated unique pieces ("bytes") that do all the work in highly differentiated ways.

The era when researchers were starting to demonstrate interval codes is mostly over. The emphasis these days is on impulse firing rate. One of the few subsequent studies supported interval coding by demonstrating that spike trains show temporal dynamics that can be identified with chaos mathematics.[13] A more recent study, based on interval-pattern detection, has also confirmed the existence of interval coding.[14]

THE BINDING PROBLEM

I have already mentioned the process in which transmitters bind to their target receptor molecule. The brain has another, more global, kind of binding that involves what I call *cognitive binding*. Actually, I coined this phrase as a result of my experiments with ambiguous figures (see chapter 4), where to perceive any one alternative perception of a vase/face illusion, for example, one has to mentally bind together the elements of the picture that could form the image of a face and another set of elements that could form the image of a vase.

This is a magical kind of process whereby the CIP contributions of perhaps millions of neurons are bound together functionally to reconstruct a complete sensation, such as a visual image. For example, in seeing a tree, all the highly limited responses of neurons scattered throughout the visual cortex reconstruct a total image. If a single neuron only detects a small edge of one leaf on a tree, how does the brain put together all that information of the bars and edges represented by the impulse trains from thousands to millions of neurons scattered all around in different places in the visual cortex? Somehow all these little representations have to get bound together to reconstruct the representation of the entire tree. Nobody knows how such reconstruction is done. It seems like a miracle.

By using CIPs flowing in multiple, parallel pathways, the brain needs to "put all together" each still frame of what the eye saw at each fixation, binding the results of processing into a coherent streaming perception and interpretation. Whether in the conscious or unconscious mind, the brain binds all these informational pieces back together to reconstitute an image. Most amazingly, the brain does this by streaming the succession of still frames in movie-like fashion in near real time.

Scientists call this the "binding problem" because nobody knows how all this disparate information is combined on the fly to re-create a holistic brain representation. The ultimate form of binding is gestalt, wherein the brain binds what it physically detects with information from memory, intuition, and imagination. Such binding is especially interesting with regard to the processing of ambiguous figures (see chapter 4).

OSCILLATION

The best guess at present is that binding has something to do with oscillation of neural populations that links activity of individual neurons into a simultaneous coupling so that their little pieces of CIP information are shared.[15] Still, this is a glib explanation, and nobody really understands it.

Reciprocal interaction among regions of mapped information should certainly facilitate perceptual binding. The maps themselves are a binding mechanism. But dynamic functional processes, such as synchronizing activity, can facilitate binding in mapped ensembles. One inevitable characteristic of mapped ensembles with feedback connections is the capacity for oscillation. Consider the diagram in figure 2.4. Those CIPs might oscillate as the pattern of activity cycles through the circuit again and again if there is no disruption from the outside, by way, for example, of the outside input at neuron number 3. Oscillatory behavior in interconnected oscillation-generating ensembles can shift phase relations of the activity in their respective populations. The sharing of information in each oscillating ensemble will certainly depend on how the oscillations are locked temporally to each other.

Oscillations of extracellular voltage rhythms, as seen in the EEG for example, are not simply epiphenomena, but reflect an essential mechanism for coordinating brain function.[16] Even so, many scientists still think of extracellular voltage fields as epiphenomena of the summed synaptic currents even though extracellular fields can modulate the impulse propagation of the very circuits that generated the fields.

Extracellular voltage rhythms arise from CIPs and can in turn engulf whole populations of nearby neurons. These rhythmic fields have the potential to synchronously modulate the excitability of many neurons such that input to the members of this population are more likely to generate impulses at specific time relations (phases) of the oscillation, while suppressing responsiveness at opposite phases. This matches the observation that impulse firing, across numerous cortical sites and behaviors, is selectively enhanced at specific phases of the ongoing oscillations created by the extracellular currents of CIPs. This is the case of the electroencephalogram, which is a mixture of oscillating frequencies arising out of the neocortex. Because electric currents have a magnetic field counterpart, there are associated oscillating magnetic fields, recorded as the magnetoencephalogram.

Frequency of the oscillations of voltages surrounding neurons (called *field potentials*, or EEG if recorded from the scalp) influence the propagation of impulses by neurons embedded within the oscillating voltage field. The external voltage fields bias whether or not a neuron in that field can generate impulses, promoting impulse generation at one point of the waveform but restricting it at the opposite peak. High-frequency oscillations allow more total impulses to be propagated. In other words, the frequency of surrounding voltage fields has the effect of chopping up time and corresponding throughput of impulses through a circuit.

Our perceptual feeling is that time moves as a smooth continuum, but to the brain time is chopped up into pieces. The duration of the pieces varies in different parts of the brain, even at the same time. Voltage-field oscillation frequency determines the duration of each successive piece of time.

Plenty of evidence exists that oscillation is a fundamental property of many brain circuits and that frequency and phase relations of oscillation correlate with the density or difficulty of information processing. The more active conscious thinking is, the higher the frequency of circuit activity in the cerebral cortex.

High-frequency oscillation is clearly associated with higher levels of conscious thought, and in my 2011 book *Atoms of Mind* I reference many neuroscientists who now use this as a guiding premise for their research. For example, if you think hard about a math problem, neocortical circuits oscillate at higher frequency and are likely to become synchronized. Oscillation can ensure that CIPs are propagated in manageable packets. The higher the frequency of oscillation, the greater the information throughput.

Notably, people can change the oscillations of their brain. Meditators can train the brain to generate alpha rhythms of eight to twelve per second or faster gamma rhythms. Buddhist monks, for example, can at will generate high frequency oscillations during their meditation.[17]

When you decide to go to sleep, your neocortex changes its dominant oscillations from high frequency to low frequency. Then, periodically during sleep, you begin to dream and the oscillations resume high-frequency operations. Of course you don't control when and how often you dream, except when you deliberately deprive yourself of sleep, in which case the brain attempts to make up lost dream time.

How precisely could simple synchronicity explain binding of visual-image fragments, for example? Synchrony generates increased activity through the routing and modulating of activity of target neural populations. This is accomplished because of the inherent electrical positive-negative cycle (equivalent to excitation and inhibition) involved with oscillation. For instance, consider two spatially distributed neural populations. When an oscillating population becomes coupled to another neural population that is oscillating at the same frequency, the oscillatory cycles will match. Since the target neurons will be receiving input during their depolarization phase, they will be easily excited. Thus, populations in the same oscillatory phases have their activity reinforced and increased whereas target populations that are out of phase will be less likely to be excited and may even neutralize each other.

The oscillatory uniting of spatially distributed neural networks may be considered the first step toward binding to form cohesive percepts. The second step is centered on the selection of such networks for further processing. Wolf Singer, a pioneer in this area of research, postulates that this selection is made as a result of the increased activity that is observed in synchronous networks.[18] Many scientists agree that synchronous oscillation is important to brain function, but nobody knows exactly what it does.

THE BRAIN'S THREE MINDS

With the discovery a little over a hundred years ago that neurons exist as distinct units, scientists realized they had taken the first step in discovering what "mind" is and how it works. If heart cells are the key to how the heart works, then nerve cells should be the key to the mind. Scientists have since learned about nerve impulses, the architecture of neural networks, and what happens biochemically and electrically at the junctions between neurons. We can now say with good assurance that thinking involves impulses, neurotransmitters, postsynaptic membrane receptors, biochemical signaling amplifiers, and postsynaptic ionic currents in multiple, parallel, and feedback circuits.[19] But it is not enough to say that these things are involved in or mediate thought. Such words are too glib, and they have little explanatory power. The knowledge accumulated over the last hundred years has, however, made it possible to hope that a true and complete theory of mind is at hand.

One reason it is so difficult to unravel the mind-brain enigma is that theorists tend to approach the problem from the wrong end. By trying to explain consciousness, for instance, theorists immediately get sidetracked from scientific issues and become trapped in their philosophical or religious premises.

In Sigmund Freud's model of the human psyche, he posited the "id" as instinctual basic drives, the "ego" as more organized cognition, and the "superego" as the consciously critical and moralizing mind. Freud deserves some of his reputation as a sex-obsessed crackpot, but he was on to something important with his id model of mind. This model is better phrased in terms of today's ideas about nonconscious, unconscious (or subconscious), and conscious mind.

Freud recognized that humans begin life as a nonconscious *id*, only later developing ego and superego as the brain develops. *Ego* emerges as a capacity for adapting the drives of the *id* to real-world realities and constraints. Much, but not all, ego operations occur with only a small element of consciousness. Many of Freud's ideas about superego are arbitrary and questionable, but fundamentally they fall within the domain of the brain's conscious mind acting on behalf of the interests of body and brain.

Rather than starting by explaining conscious mind, it may be more fruitful to take the opposite approach, beginning with operations of the simplest levels, first nonconscious mind and then unconscious mind, eventually leading to an attempt to explain conscious mind. Now it is fashionable among scholars to do away with subconscious mind and lump ideas about nonconscious and subconscious mind into a homogeneous concept of unconscious mind. However, for my purposes here, certain distinctions help our understanding.

In other words, let's think of the brain as having three minds: (1) nonconscious, which is the neural system that operates reflexively at the spinal cord and brainstem levels and is not accessible to consciousness; (2) unconscious mind, which conducts operations that are usually but not always inaccessible to consciousness; and (3) conscious mind, which conducts its operations in the bright light of explicit awareness (see figure 2.5). Despite the differences, the three minds work together as sources of knowing and doing. Reflexes, drives, emotions, awareness, intelligence, intent, and action are generated by the three embodied minds.

Nonconscious mind has very limited influence, mostly confined to reflex operations in the spinal cord and brainstem. Nonconscious mind governs our simple bodily functions, such as regulating heart rate, blood pressure, and spinal reflexes. This mind is readily explained in terms of anatomy, physiology, and biochemistry. Indeed, most of the fundamental discoveries about brain were first made in research on the nonconscious mind operations. No one speaks of this mind as "emerging" from brain function. It *is* brain function. Why wouldn't this idea apply to the other two minds?

Unconscious mind, on the other hand, influences most everything the brain does. Unconscious mind is the mind of buried memories, unrecognized desires, habits, compulsions, assorted emotions, and even a great deal of unconscious decision making. This mind operates when we sleep and operates without conscious recognition throughout our wakefulness. This mind, like nonconscious mind, can be explained by anatomy, physiology, and biochemistry. Again, no one speaks of unconscious mind as emerging from brain function. It, too, *is* brain function.

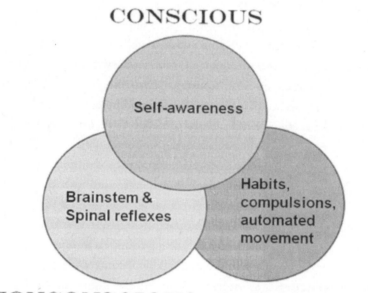

CONSCIOUS

Self-awareness

Brainstem & Spinal reflexes

Habits, compulsions, automated movement

NONCONSCIOUS   UNCONSCIOUS

Figure 2.5. The brain's three minds. The brain's total mind consists of "three minds," each of which interacts with the others. All objective evidence indicates that all these minds arise from anatomical, physiological, and biochemical processes of the embodied brain.

You may wonder why there is a distinction between nonconscious mind and unconscious mind. Both, you could argue, operate unconsciously. The distinction is subtle and perhaps not crucial to overall understanding. Unlike nonconscious mind, unconscious mind, equivalent to what is often called subconscious mind, can sometimes surface some of its operations into consciousness. It also readily accepts instruction and training from the conscious mind. You might even think of unconscious mind as conscious mind that is only temporarily unconscious—as in wakefulness succumbing to sleep.

Conscious mind has very limited functional capacity; at any one moment a person is only aware of a small fraction of all that is going on in the brain. I suspect that this happens because consciousness requires a great deal of neuronal resources just to exist, and not much is left to hold content.

Why should conscious mind be any different in principle from the other minds? Most theorists do seem to think that conscious mind is fundamentally different. Conscious mind is said to "emerge" from brain function but not be equivalent to it. The problem is that nobody knows what is really meant by saying that conscious mind is an emergent property of brain. "Emergent" just means greater than the sum of the parts. Brain scientists do, however, accept that conscious mind comes from the brain, not some out-of-body ghost.

Throughout this book my premise is that conscious mind *is* brain function. For now, let me assert that a whole mind consists of three interacting "minds," each of which *is* brain function, as is the integrated whole mind.

At this point, I need to establish some ground rules for how we should use language to describe "awareness," which is commonly used as a synonym for consciousness. This term carries a lot of anthropomorphic implication, usually inferring that awareness is a conscious operation. But people may speak, for example, of ants being aware that there is food in the kitchen. Ant brains can do many impressive things, including detecting food in the kitchen, but they are not "aware" of such a fact. When describing nonconscious and unconscious operations, we are more precise in saying that the brain "detects" stimuli and situations. To say that a person is "consciously aware" is redundant.

To illustrate the point, the nonconscious processing of the solitary tract nucleus in the brainstem enables those neurons to detect sensations from the vagus nerve with which it connects. The nucleus should not be described as being "aware" of vagus nerve input, which in turn is created by changes in our viscera. Likewise, the deep nuclei of the cerebellum detect influences from the cerebellar cortex, but they cannot be "aware" of that influence. Typically the nonconscious operations occur to regulate gut secretions and movements. Only under abnormal situations, as with indigestion, for example, will the added sensations of acid build-up, food trapped in the stomach, and such, create additional input that generates subconscious operations that will make you feel vague discomfort, though you won't know why. As indigestion progresses, the stimuli will magnify to make you consciously realize what the problem really is.

A higher level of detection can be illustrated with fear. Take fear of snakes for example. A brain area known as the amygdala generates a sense of fear upon detecting that snakes are near. Even lower animals can be afraid of snakes; in that sense they *detect* the snake stimuli. But we must ask whether they are *aware* that they have detected a snake and that snakes are something to fear? That is, do they have a conscious sense of self, and do they know that they know about snakes and why snakes are to be avoided?

Level of awareness no doubt depends on which species of animal is involved. A human, for example, surely has a different level of awareness about snakes than does a chicken.

Conscious mind cannot describe its unconscious processes, but it does know that those processes are going on and that they have consequences that might be altered—by conscious mind itself (see chapter 4). The most fundamental aspect of conscious mind is how it deals with "sense of self." That is, conscious mind knows it exists and that it is distinct from unconscious mind and able to be aware of at least some of what it knows and thinks. Moreover, things detected in the environment and internal thoughts are all viewed consciously in the context of self. It is "I" who see, "I" who hear, "I" who taste, and so on.

In conscious mind such awareness conspicuously extends across time, integrating the past and future. Some theorists like to emphasize the autobiographical nature of conscious mind, but that just refers to memory of things that happened to oneself in the past. Such integration of past and future is common among other primates, even dogs and cats. Dogs, for example, have some kind of concept of bones, which they chew on in the present. They remember that they like bones and where they hid them when they want to gnaw on them at some future time.

ON BEING AWAKE

Can you be conscious without being awake? No. Obviously, if you are not awake, you are unconsciousness. Can you be awake and yet not *completely* conscious?

Yes, as you may suspect from stumbling out of bed in the morning. Wakefulness and consciousness are not equivalent, as is demonstrated by the temporary condition of change blindness, which I will describe in chapter 4. Incomplete consciousness occurs in several disease states, such as Capgras syndrome, which is discussed in chapter 3. Another example is stroke victims with damage to the speech areas of neocortex, who can be awake while still not being conscious of speech. Another example of wakefulness without consciousness is provided by case studies of people with extensive damage of the neocortex on one side of the brain. For example, patients with damage to the right cortex, particularly in the temporal or parietal areas, may be unable to attend to or be aware of people or objects on the left side.[20] If damage is sufficiently extensive, it is as if the left side of their world has disappeared, yet they are clearly awake.

How does the brain achieve wakefulness? Typically, it is triggered by sensory stimuli (think of the morning alarm clock). All sensations (except the sense of smell) arrive into a brainstem area known as the reticular formation, as well as into specific bodily mapped targets in the thalamus and neocortex where sensations are recognized. Reticular neurons project widely into the cortex and release those neurons from inhibition (which has the effect of exciting them). It is as if the brainstem says to the cortex, "Wake up, you have information coming in that you need to recognize and process."

Brainstem arousal neurons also connect with newly discovered neurons in the hypothalamus that release a neurotransmitter known as orexin. This transmitter creates a mental tone of wakefulness. Sedatives and sleeping pills act at various points along these arousal pathways to shut them down. Orexin is of special interest these days because a new class of related sleeping pill drugs is being developed based on their ability to block orexin.[21] It is hoped that these drugs will be less addictive and have fewer side effects than traditional sedatives.

CONSCIOUSNESS IS IN THE BRAIN

As far as contemporary science can determine, there is no evidence that conscious minds are floating around in space. For all we know, each person's mind is confined to and not separable from the brain. The brain is the vessel that

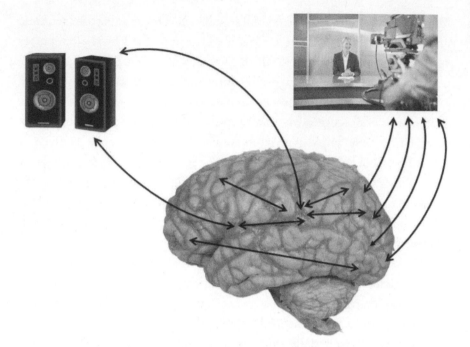

Figure 2.6. Deconstructed, distributed brain processing. When the brain receives stimuli, such as sound and images, the stimuli are deconstructed so that small parts of the stimuli register as nerve impulse discharge in different neurons. Each neuron shares this information with other regions, and some of the regions integrate different kinds of stimulus (sound and images in the example here). In this example, sound information registers in the two interconnected "speech centers," while corresponding visual information registers in the visual cortex at the back of the brain. These areas exchange information with each other, as well as with other parts of the brain (not indicated here) that might participate in supporting roles involving movement, emotion, memory, and so on. When that information is consciously recalled, CIP representations are again activated simultaneously and bound in some unknown way to reconstruct the original stimuli.

not only contains conscious mind but also generates it. Consciousness is thus not a "thing" but a brain state of being.

What part of the brain does this state come from? This is unanswerable at present, but the question is a key driver of modern neuroscience research. What we do know is this: human consciousness cannot be generated without the outer mantle of the cerebral hemispheres, the neocortex, and that mantle being made awake by the brainstem reticular formation. Some specific aspects of consciousness are controlled in specific neocortical areas, such as speech centers along the left side of the neocortex, the visual zones at the back of the neocortex, and the sound-perception area in the lateral part of the neocortex on both sides of the brain.

Humans differ from other mammals in that their conscious actions depend more on multiple areas of the neocortex. I remember once when I was a young faculty member at Iowa State University, the famous animal physiologist Hugh Dukes let me assist in a surgery in which we removed the neocortex of an anesthetized cat. After the cat recovered from anesthesia and the surgery, the cat seemed mostly normal, though obviously quite dumb. Even without a visual cortex, the cat could see and navigate around obstacles, yet it clearly had no understanding of the obstacles. Dukes used such cats to teach veterinary students about the role of the neocortex in domestic animals. But if humans suffer similar massive damage to the neocortex, they are likely to become permanent vegetables. In other words, evolution has given humans the power of the neocortex, but the price is profound dependence upon its function.

Human consciousness is enriched by the fact that most regions of the neocortex interact in processing of simultaneous stimulation with different kinds of stimuli. This is important because in the real world multiple kinds of stimuli occur simultaneously. Let me illustrate what happens to sensory input from several sources at the same time. Suppose you see a television news program and at the same time hear the news anchor speak. That information registers as CIP representations widely distributed in the respective parts of the cerebral cortex that are hardwired for vision and sound. Both kinds of stimuli are deconstructed: as Hubel and Weisel showed, images are broken down into small line segments, each selectively encoded by different neurons. A similar pitch-selective process operates for sound.[22] During registration of such information, visual information, for example, is processed in the visual cortex at the back of the head and sound information is processed in the speech centers. Multiple other areas interact (for simplicity, figure 2.6 shows only a few arrows). In short, sensory input is deconstructed, distributed, reciprocally processed, stored widely, and retrieved in a way that binds it all together to reconstruct the original stimuli.

Such deconstruction and reconstruction of information no doubt occurs at all levels of mind. But, of course, what intrigues us most is what happens when the brain becomes consciously aware of the results of the reconstruction.

What seems to be the likely source of holistic consciousness, the state itself, is functional connectivity between all parts of the neocortex and many subcortical areas that govern emotion, movements, and other functions of which we would be unaware without the neocortex. Conscious awareness of stimuli requires activation of frontal and parietal areas of the cortex. In an EEG study, researchers found that a significant difference between patients in a vegetative state and controls was an impairment of backward connectivity from frontal to temporal cortices.[23] Thus, the state of consciousness may be a top-down process that originates in the frontal cortex.

The products of conscious mind, its ideas, feelings, and thoughts, can be shared with the world outside a given brain through speech, writing, and observable deeds. In that way, many minds can contribute to the evolving nature of any given single mind as that mind experiences and learns from worldly encounters.

Many scholars are troubled by the question of how consciousness can affect the brain. Indeed, as we will consider in chapter 4, many scientists don't think consciousness does anything. They think that it makes us aware of self and some other things but that it can't cause anything. However, that is only plausible if you think of conscious mind as some sort of out-of-body "ghost in the machine," rather than itself being brain function. Paradoxically, few scientists believe in such a ghost while at the same time believing that consciousness can't affect the brain.

The problem goes away when you realize that conscious mind *is* matter. That is, mind affects matter because mind itself is matter. Once a given thought, for example, is initiated in material CIP processes of the brain, those same processes can change the brain so it can regenerate the thought and integrate it with other thoughts, past, present, and future. CIPs, obviously material in nature because they exist as electrical pulses, carry such thoughts when they are deployed in consciousness and store them long term in the form of neural circuitry that is altered in its junctions among neurons in chemical and anatomical ways that permit stored information to be reconstructed in conscious mind.

Because mind is matter, mind has the capacity to do things—even to itself! The matter is, as mentioned, CIPs. These CIPs can create change, not only in how the body moves but also in how the brain thinks. Changing how the brain thinks, if repeated, creates lasting changes in the synapses mediating such thoughts. So, here is the basis for my claim that consciousness is the brain's way to assist in programming itself (see chapter 4).

One question may never be answered by today's science: Is there such a thing as an individual soul, embedded or otherwise entangled with mind? Most people in the world believe souls exist, and this is the basis for the world's religions. The soul, by most people's definition, is not a material thing, so it makes no sense to try to explain "soul" by way of what science has revealed about the material nature of mind. This book has a focus on explaining the material basis for mind, as scientists understand it today. But in the last chapter I will mention many known material realities that I call "spooky physics," such as quantum mechanics, dark energy, dark matter, and other esoteric theories, that scientists do not understand, yet may suggest "spiritual" implications for some. "Entanglement," for example, is a verifiable aspect of quantum mechanics. Who knows: soul may also have a material existence that science is unwittingly in the process of discovering.

THE CONTEXT OF SELF

Who does the seeing of a sunset, or the hearing of a bird, or the smelling of sizzling steak? The conscious brain that sees or hears or smells does so in the context of itself. The CIP representations of these sensations operate at all three mind levels. In the early relay locations along the visual or auditory pathways, for example, the sensations are unconsciously processed in the context of self. If such stimuli are noxious, they may trigger nonconscious reflexive avoidance behavior. At the neocortex level, these same self-referenced sensations are not just merely "received," but are "perceived" consciously—again, in the context of self.

While some philosophers contend that self is an illusion, most people certainly think they have a self. Moreover, the topographical mapping of the body in the brain is unquestionably real, and the CIP representations within that mapping convince me that those representations constitute the self. Self exists as long as those CIP representations can be stored in the weightings of the synapses of the brain's global workspace and brought back "on-line," most explicitly in wakefulness. In both unconscious and conscious states, the brain knows itself in terms of its topographical body maps and the CIP representations of events and experiences that have occurred in those maps.

The global workspace idea regards brain function in terms of large-scale networks that have functionally coupled brain areas that enable consciousness and its assorted functions. Support for this idea has come from fMRI brain scans. Unfortunately, showing coincident metabolic activity in a cluster of brain areas is not equivalent to the informational connectivity mediated by CIPs. Nonetheless, brain scans have implicated a fronto-parietal network in coordinating the activity of other, more specialized networks. The notion of coordination seems to be a useful way to explain how the integration of the various subnetworks in the global workspace become recruited to orchestrate specific mental functions.

Rather than existing as some dualistic interaction between brain and mind, I regard the self simply as normal brain function. The brain generates CIPs as representations of the self and stores the representation in a way that can be unleashed from sleep or anesthesia or even sometimes coma.

Such perception is inextricably bound within a world of self and non-self. Once self is recognized consciously, the brain's inner world becomes positioned apart and distinct from the outer space, time, and events in the environment. Humans recognize self in many ways, but perhaps the most significant is our perception, often simultaneously, of our place across time—past, present, and future. Something is profoundly different when space, time, and events are processed in the light of self-conscious awareness.

How then is conscious mind different from unconscious mind? Is it generated as a special combinatorial code of CIP processes operating in multiple, dynamically changing circuit patterns? Probably. But nobody knows how this mind has an awareness process that makes it so different from unconscious mind, which also surely operates by way of CIP mechanisms, though the CIP mechanisms of unconscious mind may be coded in a different way.

A conscious sense of self emerges as a system function. The function of this system is primarily to "think" about how self can appropriately engage the environment with its own internal processing about that environment. You might want to suggest that emotions or abstract thought don't fit this model. But even here we would say that thoughts about how we feel or thoughts about ideas are really part of the way the system tries to respond to and control the environment.

A conscious sense of self is fundamental to my existence and to yours. In healthy people, the sense of self (SoS) cannot be stamped out. That is not to say that some have not tried. Among teenagers, peer pressure may smother the sense of self, but only temporarily. At a political level, Marx, Lenin, Stalin, and their followers believed that the SoS could be stamped out, and that human nature could thereby be changed to make collectivism the norm. Some communist leaders have even thought that "re-education" of one generation perpetuates genetically to create an enduring collective reminiscent of the Borg in the television series *Star Trek: The Next Generation*. Policies toward that end cannot work. They ignore biological reality.

While we do not inherit acquired characteristics, the brain does create many of its characteristics through experiences. Moreover, the real power of mind is that it gets to choose many of its experiences. Later, in chapter 4, I will explain epigenetics, which is a main mechanism by which experiences change the brain. Such processes operate in nerve cells. There is no good evidence that mental epigenetic effects occur in sex cells, and thus they can't be propagated to future generations. We can change our own brain, but we cannot genetically change the brains of our descendants.

ORIGIN OF THE SENSE OF SELF

It's dark in here. I can't see a thing. Other things are strange, too. I never feel sad or glad, nor love nor hate, nor do I have other emotions—at least in the beginning. Mostly what I do feel is the touch on my skin, and the feelings change as I move around inside whatever it is that surrounds me. Sometimes I even hurt if I get scrunched up. But I can't tell if I am being touched or am doing the touching.

I am learning that parts of me move when I feel touched. In the process I learn that I have parts that I can move. And guess what? Sometimes I can make these parts move on their own, even when they are not touched. But what is the point of moving? I have nowhere to go, nothing I can do. It's like I am in a cage.

I'll tell you what else is odd. I don't smell anything. I do taste things, but it is always the same, except sometimes a new taste comes along, but then it goes away. I hear some sounds, a regular thump-thump sound, occasionally punctuated by loud bursts of whooshing sounds. I even sometimes hear other more distant sounds.

I am beginning to sense that there are things "out there," like the tastes and sound coming from outside me. At the same time, I vaguely sense that there are things "in here," whatever that means. I am starting to develop a sense of ME.

There is not much I can do. I get sleepy and yawn a lot. My brain does go to sleep at times, and it frequently tries to dream, though I have little to dream about. My life is limited and boring, oh so boring.

You see . . . I am a fetus.

Of course, I don't know who I am yet. I don't even know what "I" means. But I have already discovered unconsciously that I exist. There seems to be ME and not ME.

Of course, I don't consciously remember these things from my experiences as a fetus. And I certainly did not have words to describe the experiences. I am just reciting what I must have experienced based on what others later told me about what life must have been like for me as a fetus.

The fetus is molded by its DNA and the physical and chemical environment of the womb into its basic features and potential capacities for achieving a final form. The fetus is developing more than a shape; it is developing its sense of self and mind. When our brains were at that stage, we didn't see anything, because after all it was dark in there. But our brains heard sounds, such as gurgling of the gut of mom, and sometimes loud noises outside our bodies. Recent research shows that a late-stage fetus is actually learning sounds that it remembers after birth. As a fetus, we tasted chemicals in the bathing fluids. We felt things, such as contractions of mom's uterine muscles. When we moved a limb, we felt the resistance it met. All these things were learning experiences, typically inscribed in mapped fashion in our brains, a place for recording sounds, another for tasting chemicals with our vomeronasal organ and tongue, and a topographical arrangement of our limbs and the sensations that came from them.

This retrospective vignette describes how sense of self is sculpted in the womb. Scholars generally recognize that sense of self is laid down in the fetus.[24] These are understandably complex biological processes. It is therefore easy for things to go awry, especially when pregnant mothers are poorly nourished or indulge in drugs or alcohol. The soon-to-be infant is at mother's mercy.

Another thing that boggles my adult mind is that I am told that all humans have many more neurons in the neocortex as a newborn than we have as adults. So whatever smarts we have acquired over the years have not come from our brain giving birth to more neurons in childhood. Though there is some equivocal evidence that a few new neurons appear in the neocortex of adults, for the most part we are born with life's quota of neurons. We better make the most of it.

In the womb, neurons divide in staged sequences by the billions from a single layer of cells. These new neurons migrate, as if they know where they are supposed to go, to their final homes in the spinal cord, brainstem, and various parts of the brain. In the process, these new neurons learn to recognize their targets, either muscles or other neurons. They thus build the unconscious sense of self, learning about their body and where everything is.

Even under normal conditions, most of these new neurons never find a target, and they die. Neurons that are not part of this sense of self have no utility. If they find a target, the target itself provides a life-sustaining force and they become incorporated into selfhood. Each surviving neuron becomes an integral part of the whole.

All this begins the process of carving out an unconscious personal identity, a self. Of course, the fetus is not consciously aware of itself, but its brain is learning that it exists and that the world exists in the forms of "in here" and "out there." The topographical maps that are being sculpted are in the process of collectively becoming a person.

Because the body is mapped most expressly in the neocortex, early scientists described the sensory and motor cortex as containing a homunculus, a "little person." That little person is a key part of a person's sense of self. Thus, the brain knows and uses these maps in the context of a self-identity. This identity can include unconscious identity, where the body registers information and commands appropriate muscle response, even if the person is not consciously aware of the whole scenario.

In late fetal stages, development moves beyond the set of circuits in the spinal cord and brainstem that regulate hormone systems and mediate basic reflexes. More complex functions appear, things like flexion and extension reflexes of limbs and reflexes involving the head, like swallowing, blinking, sneezing, gagging. Based on this it stands to reason that this fetal brain probably has an unconscious mind, one that governs basic emotions, drives, and motivations to seek pleasure and avoid discomfort or displeasure.

Late-stage fetal brains also develop signs of being awake at times. Sleep and dreaming occur in human fetuses by seven months of pregnancy.[25] Direct recording of EEG in sheep fetuses with implanted electrodes removes any doubt that the fetus can sleep and have signs of dream sleep. If this dreaming is anything like the dreaming that occurs after birth, it could indicate a rudimentary SoS deployed within the dreams where self is an observer or even participant. Dream sleep may be a crucial factor in developing conscious awareness of self. Of course, a fetus does not have much to dream about, though the activation of the brain that occurs in dreaming could be nature's way of stimulating the fetal brain to refine its topographical maps and expand network connections.

We know that in all mammals, including humans,[26] the physiological signs of dreaming occur more often in the young. Dreaming is a conscious state in which the dreamer is aware of being in a dream, either as a participant or as an observer. The bodily signs of dreaming include rapid eye movements (REM), abolished muscle tone with superimposed twitching, and an EEG that looks similar to that which occurs during alert wakefulness. When people monitored in sleep labs are awakened during such signs, they usually report that they have been dreaming.

The high incidence of REM sleep seen in late-term fetuses persists for a few years after birth. Young children exhibit a great deal of dream sleep, accounting for as much of 50 percent of sleep time, compared with about 20 percent of sleep time in adults. While it is true that young children often fail to remember dream content, I believe this reflects more a failure to remember than a failure to have conscious dreams.

We know that young children can have their conscious self engaged in dreams, as revealed by their being awakened by nightmares. Inasmuch as dreaming is a form of consciousness, this indicates that a conscious sense of self operates as a dream participant.

Self-recognition experiments that have been performed in children suggest that self-consciousness appears around age two or three. Daniel Povinelli and colleagues videotaped young children playing a game, during which an experimenter secretly put a large sticker in their hair. When shown the videotape a few minutes later, children saw themselves with the sticker, but only children older than about three reached up to their own hair to remove the sticker, demonstrating recognition that the self they saw in the video was the same as their present self.[27]

The maximum adult capacity for conscious brain function may be a moving goalpost that keeps advancing with age as long as a person continues to experience intellectual challenges. One thing we know for sure: normal brain development is not completed until individuals reach their mid-twenties. The notoriously confused judgment and behavior of many teenagers may be attributable to the fact that brain connections and function of glial support cells are incomplete.

The incompleteness of brain development also accounts for the astonishingly high incidence of mental disorders in children and teenagers. A recent survey by the United States Centers for Disease Control and Prevention found that 13 to 20 percent of American children experience mental disorders each year.[28] Most of these disorders are phobias, followed in incidence by ADHD, separation anxiety, post-traumatic stress syndrome, illicit drug use, anxiety, conduct disorders, alcohol abuse, and depression. The fact that these problems are increasing could be explained away by saying that society is just taking more notice now. I think the explanation for increasing incidence is more likely to be the negative changes in our culture. We should worry about the kind of world we are creating for our children because that world is having destructive effects on their minds. If you don't think so, you must not follow the news on television, online, or in newspapers.

WHO AND WHAT AM I?

A child begins its lifelong quest to understand who and what it is. It's one thing to say how my neurons work and what goes on when I am conscious. But what about my *being*? I know it exists, but in what form? It is not really enough to say I exist as collections of CIPs, because you could say that applies even to bugs, for which there is zero evidence of a conscious sense of self.

One thing is clear: you and I have a profound conscious sense of self. Even my unconscious brain knows "me" in terms of the bodily maps I contain and a host of stored self-referenced memories. Such a sense exists whether or not I am consciously aware of it.

Our sense of self is sculpted by the brain's experience. And it is not just what happens in the womb and as children. Even now, as adults, neurons are forming new connections all the time. New circuit possibilities abound. The nature of one's self can, and often does, change. You are not likely to be the same person now that you were as a teenager. "Thank goodness," you might say.

NEUROPLASTICITY: CHANGING MIND BY CHANGING BRAIN

There are two practical aspects of neuroplasticity: (1) environmental effects of brain development in the young, and (2) recovery from brain damage or disease. Modern social programs emphasizing early childhood experiences, such as Head Start, pre-kindergarten, and day-care programs, have been greatly affected by the behavioral research of Harry Harlow and the brain anatomy/biochemistry research of Mark Rosenzweig—and, of course, their many colleagues. I was lucky enough to know both of these pioneers. They wrote fascinating autobiographies for my book *Discovery Processes in Modern Biology*. Both of their research groups worked independently during the same era in which I "grew up" in neuroscience: the 1960s and '70s.

Harlow is most famous for his work on "surrogate mothers," though he received considerable recognition for research on learning how to learn. He did his work with monkeys. Having no lab of his own, he ended up "borrowing" monkeys from the local zoo.[29] In the early days, he focused on how monkeys learned, eventually leading to the now-famous "learning set theory." This basic phenomenon occurs in all higher animals and humans. The idea is that learning experiences develop learning templates or schemas that promote insightful capacity that expedites new learning. A recent symposium I attended revealed that there are comparable memory-forming schemas.

When Harry became interested in development of learning capacity in baby monkeys, he at first wanted to understand the influence mother nurturing has on babies. He raised monkeys under various conditions, comparing normal mother-infant nurture with monkeys separated from their mothers at a young age and raised in isolation. Comparison groups were also raised in isolation but given fake wire-mannequin mothers with nursing bottles installed in their "chests."

As babies got older, behavioral disorders developed in the monkeys raised in isolation. They were often lethargic, staring blankly for hours. Sometimes they clasped their heads, rocking their bodies back and forth. The monkeys would often respond to human intrusion by attacking themselves.

Young scientists are mostly unaware of Harlow's work, which reflects a general problem of graduate education. Namely, science is driven by grant funding that glorifies hot new research. Professor mentors are under extreme pressure to get grants. A former chancellor of my university told me, "If you can't get grants, your career is over." In this kind of pressure-cooker environment, history of science gets short shrift.

Harlow also showed that mother contact alone is not sufficient for normal development. Peer play with fellow baby monkeys was also crucial. Play is crucial for normal development of both higher animals and humans.

Rosenzweig and his colleagues performed many studies comparing the physical and chemical effects on the brains of baby rats raised in "rich" versus "poor" environmental conditions.[30] The "rich" rats were raised in groups of ten to twelve in large cages with many objects to explore and play with, and they also got some practice running mazes. Rats in a control group were raised under the standard lab conditions of three rats per bare metal wire cage with nothing to do but eat and sleep.

When the rats matured, they were sacrificed and their brains were examined. Most readily observed was the difference in brain weight, with the rich rats' brains weighing more than the brains of poor rats. The protein content of the cerebral cortex was higher in rich rats and lower in poor rats. A similar difference was seen in the amount and expression of RNA, the brain chemical that manufactures protein.

In other studies, they showed that rich environments result in adult rats that have better ability to remember new experiences and solve problems.

As for recovery from damage or disease, medical doctors have known for a long time that a cut nerve will attempt to regenerate. This is the basis for surgery that aims to reconnect a damaged nerve or nerves in a dismembered limb. Such surgery works, to some extent, but when a neuron tries to grow a new sprout into a cut nerve, that sprout has to find the tube formed by glia cells on the other side of the cut or damage. If the sprout misses the target tube, function cannot be restored. This is the problem with trying to surgically restore paraplegics. There are so many glial tubes in the spinal cord that the odds of a regenerating nerve fiber finding the right tube are next to zero.

But one aspect of plasticity provides a more hopeful note. In addition to attempts to regenerate a cut nerve, neurons in the brain grow new terminals in response to stimulation. These new terminals increase the number of synaptic connections with target neurons. This is nature's way of expanding brain capacity: the more connections, the more robust the operational capacity.

An important practical application is enhancing learning and memory. When the brain learns new things, this is equivalent to extra stimulation. Depositing this learning in a memory store requires formation of new synapses and functional circuitry. You can actually see a physical enhancement of existing synapses in electron micrographs. Such synapses have more packets of neurotransmitters and more energy-supply mitochondria, and the neuron membranes are thicker and denser.

Thus, brain cells have some capacity for repairing damage, as from trauma, blood clots or artery ruptures, drug abuse, and other causes. Consider some recent research on head trauma. Brain scans after head trauma revealed widespread decreases in blood flow. Even with local damage, the consequence typically spreads to other areas of brain that suffered no direct trauma.[31] When neurons in a given area die, for whatever cause, neighboring neurons will proliferate new sprouts into the area that is vacated by the dying neurons. They don't divide to make new neurons, they just grow new sprouts that make new connections and can enable compensating function.

Scientists are still learning how stimulation of dying neurons can trigger new sprouting, but it seems clear that certain biochemicals are responsible.

One is a protein called nerve growth factor. Another is a class of compounds called NCAMs (neuronal cell adhesion molecules). Environmental stimulation triggers the production of NCAMs, and these in turn promote formation and maintenance of new synapses. The new synapses capture the information that helped create them and make it available in the future as stored memory the next time a similar stimulus is presented. Remember, new synapses can only come from new anatomy and new biochemistry. This is a clear example of the brain building itself.

Sometimes, phenomenal recoveries from brain injury can occur. For example, significant improvement after major head trauma has been recorded. However, such improvement takes a long time and needs intense follow-up and intense mental and physical therapy. Often the prognosis is so dire that the needed follow-up is not provided. Even when it is, there are no guarantees of success. It is just impossible to predict how well treatment will work. People vary widely in their inherent genetic capacity for repair, the extent of physical damage, the appropriateness of rehabilitation strategies employed, and the diligence with which rehab is pursued.

For reasons not understood, the speed and magnitude of recovery from brain damage is affected by the patient's emotions, motivation, and active engagement in rehabilitation therapy.

These findings are meaningful for people without brain damage as well. The principles of neuroplasticity apply to development of a normal brain and its attendant mental functions. What we are was sculpted by what we thought and did. What we think and do largely determines what we become. This theme will be revisited later when we discuss the matter of free will and the question of whether consciousness does anything in chapter 4.

The adult brain contains stem cells that have the potential to divide and make new neurons. For some unknown reason this does not happen in human brains, except in the case of the hippocampus, the structure critically involved in creating explicit memories. There is some recent evidence that new neurons can appear in the neocortex, but much more research is needed to confirm and explain that.

As in the fetus, if new neurons do not get recruited into functional circuits, they will die. Thus, ongoing stimulation and brain activity is needed to help make new neurons viable. This relates to what I said earlier about epigenetic influences. A recent study has shown that individuality in behavior emerged over time in a large cohort of genetically identical mice living in a complex ("enriched") environment.[32] Individual differences appeared in exploratory activity, and this correlated with the extent of new neurons in the hippocampus. There was also a correlation with physical activity (how far each mouse moved) and how many new neurons appeared. Actually, this study was inspired from various human studies of monozygotic (or genetically identical) twins showing that they become increasingly different as they age.

One type of neuron support cell (the glia cell type that wraps around neurons to help insulate them electrically) also divides in the adult as a result of brain activity. For example, practicing a skill like juggling will increase the volume of insulated nerve fibers. Insulated neurons will die if they become denuded of their glial wrapping, and this is the fundamental problem in human multiple sclerosis (and canine distemper).

CHAPTER 3

THE NATURE OF CONSCIOUSNESS

The "Holy Grail" of neuroscience is to discover how the brain makes itself consciously aware of the feelings, ideas, intentions, decisions, and plans that are being generated in its circuits. Circuit impulse patterns are the representation of thought, and somehow conscious mind can read CIP code to know explicitly what is being represented. Consciousness has historically been treated as a kind of metaphysical, philosophical, or religious challenge, and dozens of books have been written about it. One of the philosopher giants in this area, Daniel Dennett,[1] says he has seventy-eight books in his own personal library on consciousness published before February 2004. I have a couple of shelves of such books myself.

Philosophers like to claim this topic as their turf, but philosophy provides little insight about consciousness that appeals to me as a scientist. Philosophy deals with analogies, metaphors, and allusions (even illusions and delusions!). Biologists believe there has to be a biological explanation for conscious thought (of course, biologists also have their own kinds of illusions and delusions).

We can't get very far in explaining consciousness until we know what it means to be conscious. We should start by defining consciousness. That is not a trivial task, since consciousness can be defined many ways. Certainly, consciousness includes being aware of the sensory world and our existence in it. But it is much more than that.

Perhaps one useful definition of consciousness is the state in which the brain knows it is conscious, is aware that it is aware, knows that it knows, and feels what it is feeling. One way to think about conscious awareness is that it seems to emerge from an extra set of neural processes that continues throughout a stimulus condition or memory of such a condition—and outlasts it.

Consciousness may be hard to define, but like pornography, we know it when we see it. All people know that consciousness exists in themselves, and from interacting with others, they believe strongly that others, too, are conscious. One's sense of self can never be experienced by others. The book of Proverbs (14:10) makes the point vividly: "The heart knows its own bitterness, and no stranger shares its joy." We may empathize with others, but in no way can we experience their life as they do. The consciousness of each of us is uniquely self-referenced.

Scientists and psychologists tell us that most of what the brain does occurs under the radar of consciousness. Conscious brain operations are just the tip of the iceberg of what is going on in the brain.

Figure 3.1. One way to think of consciousness. Consciousness is like the tip of an iceberg, most of which is submerged and unseen. As in the picture, the tip of consciousness sees its own reflection but does not see what is beneath. (US Air Force photo by Tech. Sgt. Dan Rea.)

Conscious awareness is performed in the context of self and non-self. You know that it is you who know. Your brain has constructed the "you" that you know from your lifetime of experiences and thought. Your brain-based identity was programmed and memorized by life experiences and conscious thoughts. The brain constructs a most intimate sense of self, and it recalls this sense of self consciously from memory after recovering from sleep.

Although consciousness can be viewed from different perspectives, I prefer to emphasize the perspective of the self. Each of us has a sense of self, and each of us engages in introspection about our self's dealings in the world. Most everything we do occurs in the context of self and non-self. One's sense of self is the core of one's being.

A related view comes from the impression that *others* have a sense of themselves, and we can infer that they think in ways similar to ours. This is the so-called Theory of Mind, which holds that every person can attribute to others mental capacities for beliefs, desires, intentions, plans, and introspection that are similar to one's own.

A third perspective is to think about how the brain uses consciousness. The brain constructs consciousness out of its own substance and uses it as a tool for more effective functioning. Consciousness is your brain's way of extending its reach into the world. Many scholars disagree, but I will challenge them in chapter 4.

To be conscious of sensations is to have mental representations of something happening outside the brain that makes the brain aware of its thought representations. Typically, we consciously recognize ongoing stimuli. Such awareness may also be indirectly generated by the brain as memories of recalled physical stimuli or experiences. The same idea applies to conscious thoughts, which typically are "heard" as voices in the head or "seen" as mental images. We can speak of the "mind's eye," the "mind's ear," the "mind's nose," and so on.

You might now ask, "What about internally generated thoughts that are not stimulus bound?" But they are stimulus bound, if not to the original stimulus, then to the memory of the original stimulus. That's why relatively little thought can be generated by babies; they don't yet have much experience to think about. This is also the basis for my dictum, "the more you know, the more you *can* know."

We can greatly extend the link to stimulus or memory of stimulus by thinking about related things or plans for future actions. The brain can do this because it has innumerable functional circuits that can be engaged to process the original information in a variety of ways and other contexts.

An important distinction has to be made between the *receiving* of sensory information ("sensation") in the brain and the *perceiving* of it in conscious mind. A good example is pain. The same stimulus that causes pain when one is conscious causes no pain when one is anesthetized.

More than fifty years ago, J. D. French and E. E. King,[2] conducted a study highly relevant to consciousness that few present-generation scientists know about. In their animal study, recording electrodes were placed at various points along a sensory pathway (spinal cord, thalamus, and the topographically mapped part of the cortex that receives input from the part of the body to which they delivered an electric shock). As shock was delivered, nerve impulses propagated along this pathway, and the summed voltages at each recording electrode reflected the induced activity. French and King compared the size and shape of the electrical response at each point when the animal was awake (presumably conscious) and when it was under anesthesia. Amazingly, the response was distinctly larger and of different shape under anesthesia.

The point is that the CIPs representing the stimulus had to have been distinctly different in the two states. The information was obviously detected in the sensory cortex during anesthesia. That is, the information was *received* but not registered in consciousness (*perceived*). The CIPs of consciousness apparently constitute a new state in which the same input information is processed in a way that enables the brain to be aware that it "knows about" the sensory input.

This study reinforces the conclusion that pain is perceived only in consciousness. Pain perception must involve more than just specific sensory pathways. Pain perception requires the global processes that are responsible for consciousness.

Perception also occurs in context of the stimulus. This is nowhere more apparent than in gestalt phenomena that possess qualities as a whole that cannot be described merely as a sum of the parts. Conscious perceptions can be distorted by ambiguous stimuli or certain drugs or by the limitations of resolution and sensitivity of the sensory receptor cells. With certain ambiguous figures, such as vase/face illusions, conscious mind actually "fills in the gaps" in sensory data by imagining sensory elements that are not even there. If one attends to the details in an ambiguous-figure image, the totality will not be perceived. But the conscious mind has a way of "backing off" to check for possible information in the image that might not be evident from detailed inspection. Conscious mind constructs meaning from limited information. This has been demonstrated from reading text in which many words have been omitted. In the case of vision, conscious mind needs only a few elements of an image to imagine the missing complimentary information. Consider figure 3.2:

Figure 3.2. Filling in the blanks. If in viewing this diagram, the so-called Kanizsa figure, you attend only to any one of the black objects, you see a "Pac-Man" object. However, it you think about all three objects at the same time, you see a white triangle covering three black circles. Of course, the triangle has no reality, but the brain uses its gestalt capacity to "invent" a new scene. This reflects the brain's capability to "fill in the blanks," as it can do when reading text where some of the words have been dropped out.

Note in the example that the brain has to have prior knowledge of what a triangle is for the Pac-Man cues to be effective. This reflects how memory works: elements of memory can be strongly associated so that recalling only a few elements will call up the memory of other, associated elements. This is yet another example of the principle that the more you know, the more you can know.

The filling-in-the-blanks phenomenon occurs in other respects. It underlies the tendency to jump to conclusions, where the mind constructs a concept, even wrongly, from limited information. It also contributes to the human tendency toward beliefs and biases, political, religious, and even scientific.

CONSCIOUS SENSE OF SELF

Neuroscience is unable at present to shed much light on the subjective experiences of consciousness. The challenge is to explain the first-person nature and functions of consciousness. To meet such challenges, we will have to understand the neural representations of the sense of self, for it is the self that has conscious experience.

Conscious realization of body mapping rises far beyond the level of topography of body parts. It creates the sense of "I." Each of my limbs is *my* limb; my lips are *my* lips; and so on. Where in the brain is this "I" that monitors much of this mapping? It surely cannot be found in any one place in the brain. I think that the "I" emerges from distributed dynamic processes, no one of which constitutes "I." the "I" must have a neural representation, which if it follows the rules of all other brain functions, is based on circuit impulse patterns.

The sense of self is bolstered in other ways, too. The hippocampus, for example, helps to form biographical memories of our conscious experiences. Many neurons in this structure also tell the brain where the body is located in space and track sequential movements through space. Thus, we remember not only what we had been doing but also where we did it.[3]

So-called mirror neurons provide yet another way the brain represents itself. Recent discoveries confirm that certain aspects of the sense of self and nonself arise from specific impulse patterns associated with when an animal acts and when it observes the same action performed by another animal or human. The neurons generating these impulse patterns are called mirror neurons. In 1996, two reports described the existence of neurons in the premotor cortex of monkeys that not only fire when the monkey makes a goal-directed act, like picking up a morsel of food, but also when the monkey sees the same act performed in a similar way by a human. Indirect evidence from brain scans indicates that mirror neurons occur also in humans.

In humans, mirror-neuron-like brain activity has been recorded from sensory, motor, and nonspecific cortical areas. For example, fMRI (functional magnetic resonance imaging) scans show increased activity in certain frontal and parietal cortex areas when a person performs an action and also when that person sees another performing the same action. Brain fMRI imaging studies reveal that contagious yawning may be mediated by mirror neurons.

A common explanation is that these mirror neurons reflect an understanding of the intentions and actions of others. Mirror neuron function has been deemed relevant to imitative learning, emotions like empathy, and even theory of mind (the recognition that other brains think like yours does). Mirror neurons likely participate in creating the sense of self because one of the fundamental features of mirror neurons is the necessity to distinguish self from non-self.

Mirror neurons help to explain how humans can promote learning just by consciously watching somebody else perform an action. This is perhaps most obvious with development of athletic skills. Athletes commonly watch films of skilled movements of others and even their own. Some sociologists say that the discovery of mirror neurons shows how humans are socially entangled, tied together in empathy and cooperative action.

One follow-up study of mirror neurons has raised the possibility that they might encode other related acts, such as behaviors that might occur subsequently.[4] For example, the consequences of a given act might differ depending on *where* the act takes place. Mirror neurons might encode differently depending on whether the act takes place within the individual's personal space, where access is more likely, or the act takes place at a distance and the observer is less likely to participate. This would test a hypothesis about understanding, because understanding of the movements could be affected by separation distance.

To test this possibility, researchers recorded mirror neurons when an object was within a monkey's personal space (within arm's reach) and when it was some twenty-eight centimeters or more away.[5] Movement-related neurons were activated by both the execution of the required movement or by observing a human make the same movements. But 53 percent of the mirror neurons fired selectively, depending on where the object was located relative to the monkey's personal space. These space-sensitive neurons were about equally divided between those that were preferentially responsive when the object was in the monkey's personal space and those that responded when the experimenter manipulated the object's location. Thus, it would seem not only that mirror neurons encode the movements, but also that a subset of them encodes the spatial context of the act and an understanding of the implications, such as "mine" and "not mine." That is, this subset could be part of the brain's self-representation.

The investigators concluded that these space-sensitive neurons are important for evaluating subsequent interacting behaviors. That is, the only way a monkey can interact when the object is outside its personal space is to plan movements to get closer to the object. If the object is in the personal space, then competition for the object might be anticipated. These neurons might be trying to answer the question, "How might I interact with the experimenter?"

An alternative explanation has profound implications. Space-sensitive neurons could be coding for the object's relevance to the monkey's sense of self. Those mirror neurons that fire selectively when the object is within personal space may be part of a larger circuit that contains the sense-of-self representation. Close objects are viewed as a component of the sense of self (for example, "This object is or can be mine"). Mirror neurons that are selective for objects outside of personal space suggest that the object might belong to others ("It's not mine, at least not yet"). A vivid way to illustrate the difference is the difference between the possessiveness a tethered dog shows for a bone placed at its feet and a bone placed beyond its reach.

What these ideas also suggest is that a conscious concept of "me" is encoded differently from a concept of "mine." In both cases, the concept is represented by circuit-specific patterns of impulses.

The idea of mirror neurons can be extended beyond the realm of movements. By observing the facial expressions and body language of others, we can understand their emotional expressions. This being the case, mirror neurons must play a role in theory of mind and social behavior.

The conscious sense of self (SoS) is the most profound human sense. Its formation begins in the womb as the developing brain discovers its body parts and that there is an "in here" and an "out there." As we mature and SoS becomes explicit in consciousness, the brain learns to value SoS above most everything else. It responsible for the compelling human need to be valued, to love, and to be loved. It is the basis for self-confidence and self-esteem.

Ever wonder whether animals know they exist in a conscious way? To illustrate the question, consider animals looking in a mirror. Most species, if they will look in a mirror at all, may think that what they see is another animal. Higher animals, like dogs, soon realize what they see is not another real dog (to a dog, another dog is something that smells or sounds like a dog). Animals looking in a mirror do not seem to make much mental connection with the image in the mirror. The reason may be that vision is not their primary sensory cue for sense of self (as smell is to dogs, or sound is to other animals).

It is important to know how important neocortical sensory maps are to sensory processing and that the number of neurons committed to specific body parts varies with species.[6] For example, the pig has a disproportionate amount of circuitry devoted to the mouth and snout.[7] Humans have disproportionate representation of lips and fingers. Thus, it seems likely that the sensory cues we most rely on for conscious sense of self may vary by species, depending on the extensiveness of the representation of various body parts. A pig may be more consciously aware of the garbage it roots around in than of other sources of sensation.

In general, I think the reason that animals don't recognize themselves in mirrors is that mirrors don't provide the right kind of cues. Maybe a dog would recognize itself in a mirror if the mirror provided dog odor. In the case of primates, you can get them to look in a mirror, and the more advanced species even seem to recognize themselves, especially when they touch themselves and feel the sensations at the same time they see corresponding movement in the mirror.

A famous experiment was reported in 1970 by my old rival, Gordon Gallup, in another line of research ("animal hypnosis").[8] He allowed chimps to play with a mirror for ten days. Then he put a colored mark on their foreheads while they were anesthetized. They paid no attention to the mark until they were given a mirror again. Then they would touch their foreheads, using the mirror to guide their hands as they touched the marks. Gallup interpreted this to indicate that the chimps were recognizing themselves in the mirror.[9] Since that time, such self-recognition has been observed in great ape species but not in monkeys.

Gallup tried to extend this observation to suggest that apes are like humans in the sense that both have a theory of mind; that is, both humans and apes know how their own minds work, which allows them to have a theory about how others' minds work. Some neuroscientists reject Gallup's notion that apes have theory-of-mind capability, but they do accept that mirror self-recognition is a necessary first step in consciousness development.

Although dogs don't recognize themselves in a mirror, they do recognize their names and clearly seem to know that when their name is called, it refers to them. Dogs can even plan, as indicated by the hiding of bones. My dog Zoe will even bring a bone she has hidden in the field onto the porch in early evening on her way into the house so she will have it available later when she is put out for the night.

This book is not the place to debate animal sentience. But that possibility does suggest there could be a continuum of consciousness capability in higher mammals that is paralleled by evolutionary development of the brain, particularly the neocortex, which is the seat of consciousness. In that connection, if you have ever dissected the brains of dogs or horses or cows or even sheep, you know that these are complex, sophisticated brains, built along the same lines as the human brain. They, too, have an impressive-looking neocortex, though it's much smaller than the human brain. Such animals may not be as stupid as they seem. Lacking hands and speech, they are severely constrained in ability to express whatever intelligence they have. When intelligence tests are tailored to fit the behavioral repertoire of a given animal species, the animals often show that they are much more intelligent than you might think. Pigs, for example, stand out in this regard, which did not surprise me, as I had raised pigs when I was a teenager and seen them do surprising things.

A sense of self is necessary, but not sufficient for full consciousness. Introspection about one's self is the hallmark of higher consciousness. A dog, for example, clearly has a sense of self, but that does not mean it engages in introspection about what it means to be a dog or its own dog self.

A few humans have an unfortunate disease in which they cannot recognize themselves, and they often believe that loved ones or even pets and objects have been replaced by impostors. This disease is known as Capgras syndrome, named after the French psychiatrist who first recognized it. Looking in the mirror, a Capgras sufferer named John might shave around his mustache. The man in the mirror shaves around the mustache. But John does not make the connection. That is not John in the mirror. If he shaves off the mustache, he notices that the guy in the mirror did also without recognizing that he, John, is the guy in the mirror.

Capgras patients have a cognitive disconnect. What they see in the mirror is not interpreted as a representation of their own persona, but rather as that of someone else, an imposter. This may originate in a deep-seated unconscious psychosis in which patients may have severely negative emotions about themselves. Sometimes antipsychotic drugs can make this delusion go away. Sometimes cognitive therapy works, wherein patients are systematically instructed in small steps to learn the causes of their misinterpretation. Note that this is the indication of the teaching function of consciousness, which is a recurrent theme of this book.

MASS ACTION, BRAIN SIZE, AND CONSCIOUSNESS

Recall the earlier conclusion that consciousness depends on brain organization and that the smaller brains of lower animals may only allow them to be conscious of some things and oblivious to the rest. It follows that the larger brain of humans would support the highest level of consciousness capability. Some scholars think that consciousness is the product of increased brain size, especially size of the cerebral cortex. But brain size does not apply to consciousness as much as does brain organization.

The "size-matters" view arises from the common observation that animals with the biggest brains, relative to the body mass that those brains must innervate, are the smartest animals. Examples include elephants, porpoises, apes, chimpanzees, and, of course, humans. Most people extend this observation to attribute at least some level of consciousness to big-brained animals. Yet intelligence and consciousness are not the same thing.

The size bias was buttressed by a most scholarly and significant book by Walter Freeman in 1975, *Mass Action in the Nervous System*. Freeman's well-documented theme was that the advanced functions of the brain arise from collaborative interaction of large pools of neurons. Although not the point of Freeman's book, it would seem logical that more pools of neurons results in a greater degree of consciousness.

Though seldom recognized, "packing density" of neuron fibers is another factor relevant to intelligence. In mammals, for example, small brains have to be jammed into small skulls—as in mice, where neurons and their processes are crammed together inside that small skull. Larger rodents, like rats, are smarter in part because their brains reside in a bigger skull and their packing density is less. Why does that matter? Larger skulls permit growth of new neural outgrowths in response to learning experiences and formation of memory. A densely packed brain has reached its limit for physical expansion of neuronal processes.

A basis for overturning this brain-size bias, as far as consciousness is concerned, was established inadvertently with the work begun in 1952 by Roger Sperry,[10] which led to his Nobel Prize. He and his graduate students originally studied what happens to the thinking ability of cats if you separate the two cerebral hemispheres by surgically cutting the huge connecting fiber tract known as the corpus callosum. They demonstrated that while such cats seemed behaviorally normal, they could be trained to perform opposite choices with the two eyes, each of which only communicated with one of the hemispheres. However, since this was done in cats without any testing of consciousness, as such, they could not speculate about how consciousness was affected.

Sperry soon conducted studies on split-brain monkeys, and the idea began to emerge that each surgically isolated hemisphere could generate its own operations. Soon thereafter, neurosurgeon Joseph Bogen became aware of Sperry's work and approached him with the suggestion that it might be safe to cut the corpus callosum of human epileptic patients whose disease was so severe as to be life threatening. The idea was that spread of a seizure could be interrupted by cutting this communication cable between the hemispheres. Eventually, surgeons began to perform this helpful procedure, with the patients later assigned to Sperry for testing of mental functions.

Human split-brain conscious functions were readily apparent. If verbal messages were sent via the left ear to the left hemisphere, the patients understood, but no understanding occurred when messages were sent via the right ear. This confirmed that conscious processing of language is normally a left-hemisphere function. In other words, you only need half a brain to consciously use language. This left open the question of what, if anything, the right hemisphere does.

By clever manipulations of how visual information was presented to the eyes, Sperry and his students showed that conscious functions do occur in the isolated right hemisphere, again being performed consciously by only half a brain. These typically involved recognition and manipulation of geometric figures. They even showed some language processing in the right hemisphere, for example, by projecting a picture of a name of a common object to the right hemisphere and instructing the patient to retrieve the corresponding object by hand or by pointing to it. Not markedly disrupted in split-brain patients were the subconscious emotions normally associated with certain tasks and objects.

Neither Sperry nor those who followed similar lines of research made much of the fact that the tasks separately performed by each hemisphere were mediated in consciousness without any help from the other hemisphere. Such patients could be conscious even with "half a brain tied behind their back." Actually, they could operate two separate conscious minds simultaneously using half a brain each.

Thus, it seems that consciousness, as a state, depends more on circuit organization than on the total number of circuits. The range of things one can be conscious of probably does depend on the number of circuits, but apparently consciousness itself does not. This is an important clue that researchers should pay more attention to neural-circuit design.

From the earlier explanation of cortical columns and their interconnections, we should conclude that it is the coordination of activity among such columns that gives rise to consciousness. This may help explain how elderly humans who have lost a substantial amount of cortex from disease may still remain conscious, though of course with diminished mental capability. Likewise, we might conclude that animal species don't have to have big brains to have limited consciousness, as long as they have highly interconnected cortical columns.

CAUSES OF CONSCIOUSNESS

Brainstem Trigger of Wakefulness

Wakefulness does not just happen. The CIPs that represent a state of wakefulness are driven into being by activity of a cluster of neurons in the brainstem. Back in the 1950s, a lot of excitement was generated by discoveries that implicated parts of the brainstem in consciousness.[11] Specifically, the central core of the brainstem was revealed to be essential for driving the emergence of consciousness. Delivering mild electrical current to these cells via surgically implanted electrodes in sleeping cats aroused them from sleep. The same electrical stimulation elsewhere in the brain had no such effect. Under normal conditions, these same cells are directly activated by all sensory input (except odors), and these neurons activate all regions of the cerebral cortex to produce the associated wakefulness. So while sensory information is arriving in the topographically mapped sensory neocortex, the rest of the cortex is getting the message to "wake up and process what is coming in."

Damage to this brainstem area causes permanent coma. Stimulating this same area in a normal person increases the person's conscious alertness or wakes him or her up from a sleep state. These original reports were stunning. One of their impacts was to inspire me and many others to pursue neuroscience. The fact that such studies never led to a Nobel Prize is one of history's great academic injustices.

Consciousness, though typically triggered and sustained from the brainstem, operates through the circuits of an aroused neocortex. Clusters of certain neurons in the basal forebrain and thalamus are also involved. Collectively, these areas form a consciousness system. No one part alone can produce consciousness. The brainstem cannot sustain consciousness because it lacks the complex cortical-column network architecture of the neocortex.

Knowing what triggers consciousness does not explain the neural nature of consciousness. Neuroscientists generally approach the issue with the common view that mind is constructed by well-known neural processes, such as registering sensory input, comparing it with stored memories in terms of content and emotional impact, and making "decisions" about the appropriate interpretations and responses. This mind may operate below the level of consciousness, so we are left with the enigma of explaining the difference between conscious and unconscious. Neuroscientists call this "the hard problem" (see below).

Conscious mind emerges when the consequences of neural processes are expressed explicitly, perhaps re-created in terms of sounds, sights, or other sensations that are associated with the events triggering consciousness. For example, if I hear a gunshot outside my door, my unconscious mind registers the sound, detects the location in space from which it came, and compares that kind of sound with other kinds of sound that the subconscious mind has experienced and remembered. My conscious mind provides the brain with a way to know what the brain is doing at the present moment and to adjust and respond to the sound processing. That is, I know that my brain has determined the sound originated outside my door and that the sound is from a gun. Simultaneously, I may also become aware of my memory that guns can be dangerous, that there are bad people out there who shoot people, and that I may therefore need to call the police and make some attempt to protect myself. Or if it is dove-hunting season, my conscious mind will tell me the sound came from a hunter.

To unravel "the hard problem," I think it will be fruitful to begin with operations at the simplest levels, first nonconscious mind and then unconscious mind. My stance is that by studying CIPs and oscillatory circuitry, we will come to see that mind is not a ghost, but is matter.

In humans, the most basic mind from an evolutionary perspective exists as nonconscious reflexes and controls over visceral functions. What we know about nonconscious mind is very well established. Much of the nonconscious mind emanates from the brainstem and its peripheral connections.[12] We know that information in nonconscious mind exists as CIPs.

When we think consciously about things we experience, there is a CIP representation attributed to the relevant senses: smell, taste, touch, sight, or sound. Higher-level thought, such as that engaged by music, language, or mathematics, is also based on CIP representations experienced in the context of self. Disembodied mind is not possible—at least not in the four-dimensional universe that we experience.

All three kinds of mind must operate with CIPs. Obviously something must be different about the CIPs of consciousness. What might that be?

First, let us realize that consciousness depends on the neocortex. For example, if disease or trauma destroys the visual cortex at the back of the head, conscious visual sensations are lost. If a stroke destroys the language centers in the cortex, language ability is lost. Damage anywhere in the cortex perturbs the normal CIPs and usually produces some abnormality in conscious function.

Learning How to Be Conscious

Recall that wakefulness is necessary but not sufficient for consciousness. We are still left with the question of the biological nature of consciousness. I mean *consciousness* in a deeper sense than the ability of brain to generate conscious awareness of things that don't exist, as in the Pac-Man example above. Even so, think about the point that conscious realization of a Pac-Man can only occur if you know what a Pac-Man is. I will return to this point after dispelling the common notion that language is what makes us conscious.

Some scholars think that only humans have language, that language causes consciousness, and therefore that only humans are conscious. But both of the premises are wrong. Animals lack language, but they communicate with sounds (elephants, birds, dogs, even cows), not to mention extensive body language.

But this is not the syntax-based symbolic communication used by humans. Even so, some animal species act as if they have conscious awareness of themselves and the ability to distinguish themselves from others. Don't think so? Then explain why certain animals have personal space that they extend to larger territories. Watch an animal, like a dog or a cat, groom itself. A dog has personal possessions, such as toys and bones (and owners). Cats possess things, too (including their owners). Though such things are not proof of consciousness, they are certainly suggestive. At a minimum, such animals have a clear sense of self and non-self.

I have already produced one proof that language is not a cause of consciousness. Clearly, you don't have to use words to perceive the nonexistent triangle in figure 3.2. Your consciousness pre-exists and is independent of whether or not you use the word *triangle* when perceiving it.

Language does greatly enrich consciousness, of course, but it is not essential for it. In fact, some forms of transcendental meditation shut down silent chatter of the mind with a conscious focus on breathing, yielding an awareness that has no language content. Artists, especially modern artists, rarely use language to guide their artistic creations. A football quarterback does not use language to "read" a defense. All of us consciously conceive and interpret images without accompanying language. We can be aware of our emotions and the varied input of our senses even when we can't explain these things in words. If you still find such anecdotes unsatisfying, consider the documented evidence that people with lesions of cortical speech centers may no longer have language capability yet are still conscious. Then there is the consciousness of the right hemisphere we discussed earlier in split-brain patients.

One thing nobody seems to write about is the conviction I have that the human conscious sense of self does not automatically appear. The brain *learns* it, and it probably begins learning it as a late-stage fetus. As mentioned earlier, this learning begins in the womb, as the emerging brain creates the neural mapping of its body, both in terms of where stimulation of body parts is coming from and in terms of how the brain can make specific body parts move in controlled ways. This is learning about self at its most fundamental level, and this learning is prerequisite for learning sense of self consciously.

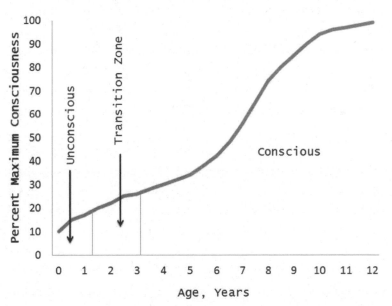

Figure 3.3. Learning the sense of self. Sense of self begins in the womb as body maps are generated. In early childhood, the sense of self gradually transitions into a conscious awareness of oneself. This self-awareness process refines and grows as the brain develops and life experience expands. Though the graph depicts the common idea that puberty culminates the self-awareness process, older people know from their own experience that the content of such awareness can change throughout life.

Learning a sense of self continues throughout life. In early childhood, as mentioned, a child may not have much of a conscious sense of self. But we know from watching children grow, and from remembering our own life experience, that consciousness becomes more robust and profound as we age. Some religious traditions hold that conscious maturation occurs at puberty, as suggested in figure 3.3, but more subtle development surely occurs earlier and throughout a life span.

CIP Basis of Consciousness

Research over the last hundred years has explained the general principles of how nervous systems work. Most of that research was originally performed on experimental animals, and the findings have been supported by observations in humans. I, like most brain researchers, performed a great bulk of my research on animals because it is often not practical to experiment on humans. But the discoveries generated from animal experimentation help us to understand the human brain. Most of these insights help explain nonconscious mind. Why throw all that understanding out in our modern efforts to understand subconscious and conscious mind? Could they not operate as extensions of the same basic principles?

Many pet owners will swear that their pets are conscious. Dogs, for example, most certainly have a conscious sense of self. Of course their consciousness level cannot equal ours. But here is an anecdotal bone you might chew on: I live in the country where many dogs, including mine, roam freely. A large and mean neighbor dog would daily invade my dog's territory and start fights. After my dog won a few of the fights, the neighbor dog still came part way into my dog's territory but did not start fights any more. My dog had drawn an imaginary line in the field, where my dog and the neighbor dog would face off at roughly the same time every afternoon. Both knew where that line was (the line never moved) and both chose not to cross it. They would just sit on their respective side of the line and glare at each other. In the afternoons, before the "appointed" face-off time, my dog would go early to the face-off line and wait for the neighbor dog to show up (which he usually did). The neighbor dog caused so much trouble with other dogs that somebody shot him.

Study of mental behavior in animals has provided compelling evidence that the currency of "thought" is the nerve impulse, enabled by biochemistry and circuit anatomy. Likewise, I would expect the currency of unconscious and conscious mind to also be the nerve impulse, or, more precisely, the spatial and temporal patterns of impulses in distributed and linked microcircuits such as the cortical columns.

There are, of course, many biochemical and physiological phenomena associated with nerve impulses. These give rise to a wide range of correlates of consciousness. But correlates are not always necessary or sufficient to explain consciousness.[13] Because the currency of thought is the nerve impulse, the currency of conscious thought must also be circuit impulse patterns, most likely a special way that cortical-column CIPs are linked in time.

To make sure we don't look for consciousness in "all the wrong places," I promote the view that research should focus on CIPs and the phase-relation binding of electrical activity among cortical circuits. A starting point of explanation can be how sensory stimuli are registered in the brain. We know from monitoring known anatomical pathways for specific sensations that the brain abstracts elements of the outside world to create CIP representations. As long as the CIPs remain active in real time, the representation of sensation is intact and may even be accessible to consciousness. However, if something disrupts ongoing CIPs to create a different set of CIPs, as would happen with a different stimulus, the original representation disappears and may be lost. Of course, the original CIPs may have been sustained long enough to have been consolidated in memory, in which case retrieval back into active working memory could reconstruct the CIP representation of the original stimulus.

How, then, can we relate these facts to the issue of consciousness? Consider the possibility that conscious mind has its own CIP representation for its sense of self. This would suggest that consciousness and sense of self are normally conflated, explaining how the CIP representation of consciousness makes one conscious of self. Because the brain is aware of self in the unconscious state, we might presume that what changes in the conscious state is consciousness itself. Specifically, when the brain constructs a sense of self, it must do so in terms of neural representation, which I argue takes the form of CIPs. That representation of self is preserved when the CIPs of consciousness are launched to access the sense explicitly.

Let us now think of consciousness in terms of function of a metacircuit that integrates CIPs from multiple contributing circuits—like cortical columns. Because consciousness requires coordinated activity in widely distributed cortical circuits, we can suspect that consciousness is launched whenever a certain set of time-locked CIPs occurs in the metacircuit.

How can time locking create consciousness? No one knows, but some possibilities seem intuitive. Time locking of multiple CIPs produces shared information across the participating circuits. As a result, the total information representation in the aggregate metacircuit is much greater. This enriched information content includes not only that of the separate CIPs but also that from the accumulated CIPs. Perhaps that extra information is what enables awareness of being aware.

In any case, consciousness would seem to be *constructed* rather than emergent, as most neuroscientists believe. Thus, capacity for consciousness and the conscious sense of self-identity can grow with time, being modified by biological maturation, learning experience, and recalled memory of the stored sense of self.

This view may also explain how the "ghost in the machine" can do things, for if conscious mind is really a special CIP state, those CIPs can make things happen, just as nonconscious and unconscious CIPs do. The upshot of it all is that the conscious mind can act as the brain's active agent, much like an avatar partner that operates in parallel and in conjunction with unconscious mind to make total brain function more adaptive and powerful than could be achieved with unconscious mind only. Evolutionarily, this may be the mechanism that changed prehuman zombies into who we are today.

It is useful to think of certain circuits or parts of certain brain areas (most likely in the cortical columns) as acting as network nodes that functionally hold together the circuits within each immediate domain and act as relay and communication points with other neural hubs. This idea is analogous to the way commercial airline hubs operate. A perturbation of any one node has the possibility of spreading to other nodes and their circuit domains. If enough of these nodes become temporally linked, consciousness and coherent thought may emerge. Perhaps consciousness emerges when these hubs become extensively synched at certain frequencies.

Regardless of which CIPs produce the "I" of consciousness, those processes should also be capable of willed modification of that processing according to the nature of their output, which is represented in the consciousness. Remember, consciousness and sense of self are conflated. When we have a conscious experience, the neural processes that make us aware of the output of those circuits provide a physical substrate for a person to make an adjustment, which may also be manifest in the consciousness. When I say to myself, "stop smoking," my unconscious brain circuits that generated the desire to smoke also made me aware that I want to smoke. But other brain circuits have been programmed, by information that my brain had received and made manifest in my consciousness, to resist smoking because it is unhealthy. Conscious activation of the CIP representations for the reasons not to smoke should suppress activity in the circuits that would otherwise make me pick up a cigarette. In other words, the brain can control its own processes.

It is not just mind over matter. Mind IS *matter.*

Synchronization

Now we come to the crucial question of how conscious mind can "read" what is in working memory and deploy it in a conscious "thought engine" (see "How We Think When Conscious" below). Consciousness could be thought of as being created from a constellation of neuronal assemblies that becomes sufficiently coordinated so that each component circuit "knows" what other circuits are doing. This capacity may emerge from reaching the threshold of a critical mass of circuitry that has become phase locked at certain frequencies.

It is now abundantly clear that conscious states are supported, and perhaps created by, phase locking of activity from multiple distributed populations. To date, no experimenter has tested questions involving what happens to phase locking of key brain areas in the same individual brain during various transition states along the consciousness continuum. But numerous studies of a single state indicate that oscillatory synchronization is a crucial component. For example, when the conscious mind is grappling with a difficult problem, many areas of the cerebral cortex generate high-frequency ("gamma") brain waves. Across multiple neuron assemblies, these oscillations are often synchronized in high-intensity thought.

Figure 3.4. Example ambiguous figures. Diagrams like these are ambiguous in that they can be perceived in two ways. Though the visual stimuli in any one image remains constant, we consciously think of it one way or another. Your brain sees what it chooses to see from its stored memory of a vase or two faces in profile (*top*), a man's face or a naked lady (*left*), or an old woman or a young lady (*right*).

An experiment conducted in my lab demonstrated that oscillatory coherence changes during the transition between unconscious and conscious realization of ambiguous-figure stimuli (see figure 3.4).[14] Note that for a given image, the stimulus impinging on the eye is constant. It is the conscious percept that changes.

THE "HARD PROBLEM"

Beyond conscious sense of self, consciousness allows us to experience life explicitly. The "hard problem" as scholars put it, is determining how the brain allows some things to be processed explicitly. Clearly, there is something biologically unique about consciousness.

For centuries philosophers and scientists have grappled with what has been called the "mind-brain problem." In the nineteenth century, people thought of conscious mind as a "ghost in the machine," and perhaps most people regard it that way today. That is, people accept that consciousness seems to come from the brain, but it has a ghost-like quality and may not seem to be a physical entity. Consciousness seems inextricably linked with vague notions of spirit or soul. By the twentieth century, science began to show that conscious mind might have a material basis. In the twenty-first century, science may be able to explain that material basis.

All modern theories assume the evolutionary requirement for lots of new cortical circuitry to support consciousness as an added dimension to the sense of self possessed by higher animals and humans. None of these theories place sufficient emphasis, in my view, on the essential nature of cortical-column network architecture. Recall that each cortical column has local processing capability, yet because of distributed inputs and outputs each column is embedded with other columns in more global metacircuit operations. The extended cortical circuitry idea, sometimes referred to as the "Global Neuronal Workspace," holds that many of the cortical neurons across all parts of the cortex are widely interconnected and collectively constitute a global workspace. There is certainly plenty of anatomical evidence that our neocortex is built this way. Anyway, the workspace idea holds that localized networks of neuronal processes, such as might be going on in the emotional part of my brain, or in the auditory part, or the visual part, and so on, compete for attention. That is, these multiple parallel operations compete for access to the global workspace.

This architecture affects the electrical functions of the neocortex, which are often monitored using the electroencephalogram (EEG). A comparable index of activity of the global workspace can be accomplished by recording magnetic fields from the scalp, but that is technically more complicated (and the equipment is much more expensive). Global workspace operations tend to manifest as relatively slow alpha and beta rhythms, whereas local operations tend to be reflected in higher frequency (>30/sec) gamma oscillations. The power and phase of slow rhythms, like alpha rhythms, are much more likely to synchronize over large areas of cortex and be modulated by a wide variety of different thinking processes, such as attention and working memory.[15] All these rhythms exist during consciousness, but their magnitude and synchronization vary widely depending on what the brain is sensing and thinking. For example, most people generate many alpha rhythms when they are relaxed and have their eyes closed. But these alpha rhythms go away when our eyes open, presumably because many local workspaces must then process the multiple features of the visual stimuli we begin to receive. In effect these local workspaces combine to provide bottom-up modulation of the global workspace in which they are embedded.

Consciousness as such is likely a more global process. The conscious sense of self could be a "top down" process, where conscious mind "looks in" on local processes and selects which ones it wants to pay attention to, perhaps recruiting local workspaces as the thinking situation requires.

Well, there is an ancient idea like this called the *Cartesian Theatre*. It is named after René Descartes (1596–1650), a French philosopher. In modern times, this notion was made more scientific by the suggestion that the neocortex's sensory and motor maps provide a representation of a little person inside the brain. The activity going on in the various local neural networks, as described for the global workspace idea, is presented as episodes on the stage of a "Cartesian Theatre" for viewing by the virtual little person.

The normal approach for understanding consciousness is to think of it as a state in which the impulse patterns of routine thought become explicitly accessible. That is, when conscious, the brain can read many of its impulse representations.

But suppose it is the other way around: the brain becomes conscious when its impulse patterns become bundled in small chunks of working memory. The circuitry processing such chunks can focus neural resources intensely, achieving conscious realization, because the information streams are bundled in manageable small chunks. Such chunking could readily be achieved by high-frequency oscillation, particularly if such oscillating activity becomes time locked across the multiple neuronal ensembles in the global workspace.

Both the global-workspace and the Cartesian-Theatre ideas emphasize a limit on the information that can be accessed or viewed in consciousness. Ultimate information capacity is held to be the sum of all the local cortical-column workspaces. This accords with our common experience that conscious working memory has very limited capacity. At any one moment, most people can only hold four independent items of information in working memory.[16]

So how then do we think, create new ideas, or solve problems? The information in our working memory must stream in, one small batch at a time, in a rather seamless way to our "thought engine," which is basically the collection of local workspaces that accomplish the various facets of thinking.

CONSCIOUS THOUGHT

Who we are is largely learned through experience, and much of this learning has occurred and is "remembered" implicitly and unconsciously.[17] These representations may not be accessible consciously, but they can affect our behavior. Though this book later presents the argument that people have free will, it is nonetheless true that without consciously exerting introspection and free will, our thoughts and behaviors are driven by the unconscious mind.

As claimed at the outset, the construction of the sense of self begins in the womb, and the SoS is then modified through life experience. You might say that we make it up as we go along. If we thought more often about this, we might be more careful about what we think and do, for the consequences of our thoughts and actions extend beyond their immediate outcome; they also contribute to our ongoing self-construction (or destruction, as the case may be). Even though this construction of the self is most evident during childhood, it is a life-long process.

Brain anatomy continues to change in obvious ways up until about the age of twenty-five. The last visible change occurs is in the prefrontal cortex, where we weigh alternatives, make judgments, plan for the future, monitor our behavior, and exert executive control. In addition, microscopic and biochemical changes continue into old age, beyond the age at which gross anatomical change can be detected.

How We Think When Conscious

Thinking requires the mind to operate in small iterative steps: recognize what is being thought about, make an adjustment, such as choice or decision, recognize the changed thought, and then perhaps progress into action of some sort. This would mean that conscious thinking operates as a succession of small time frames, played out like a movie in which the contents of a given frame can be changed as it goes along. Perhaps this is most evident when a person delivers a monologue. What is in our consciousness at any given moment can be something remembered or something currently being experienced. In either case, we hold these representations as short snapshots in a narrow window of time lasting only a few seconds, flitting from frame to frame in the stream of consciousness. The chopped-up time appears seamless because the transitions between thought segments are so rapid—like still frames in a movie. Analytical thinking or problem solving can occur when we string together these snapshots in a coherent and systematic way.

The speed at which we think consciously can be thought of as a frame rate. Frame rate may be determined by the time chopping that results from oscillation. Just as "working memory" has a limited capacity, our "working consciousness" frames are likewise built upon limited spans of time. One of the implications of this operational mode for conscious thought is that for consciousness to do anything constructive, such as reaching a complex decision or solving a problem, conscious processes must guide the process to keep it on track. In making a decision, for example, I must consciously orchestrate the process whereby each snapshot of alternative choices and their consequences is viewed, organized, evaluated, and then remembered before a final choice can be made. Just as formation of memory can be interfered with by disruptions that occur, our stream of consciousness can likewise be disrupted by interjection of incompatible or irrelevant frames of consciousness. Limited working-consciousness span leads to what we call mind wandering or loss of focus.

Working Memory Biology

Conscious thinking spans past, present, and future. Could you think consciously without working memory? No. Working memory is the capacity to hold a limited amount of information (such as a telephone number) in conscious awareness long enough to make use of that information (as in actual dialing of the number).[18] A helpful metaphor might be the cellophane writing pad you may have played with as a child, where lifting the transparency erases what was written and provides a blank sheet for writing something new.

Any theory of conscious thought needs to accommodate working memory, because conscious thought is not possible without it. When we are thinking consciously, our thoughts are held in the "real time" of working memory. This fact presents two theoretical questions: (1) what is the biological basis for working memory? and (2) how does the brain make itself aware of what is going on in the circuits that support working memory?

All kinds of thought progress by shuttling information on to and off of the working memory "cellophane writing pad." Working memory during thinking can be viewed as a place holder for a succession of the elements of thinking. If you are consciously thinking about solving a math problem, for example, each step is successively brought into working memory, used as input for a "thought engine" that uses the succession of input from working memory to solve the problem (see figure 3.5). The thought elements, working memory and the "thought engine," operate via CIPs that are neural representations of the respective information.

In solving a math problem of several steps, for example, you have to be consciously aware of each step, but you work out the problem sequentially and do not perform all steps at once. The CIP representation for each step shuttles onto the notepad, either from memory or current instruction, and then is passed into the processing thought that prepares for the next batch of working memory CIPs.

Where is the notepad? This virtual notepad probably is a whole set of circuits in which there is considerable overlap with those circuits that supply input and those in the "thought engine" that act on the input.

There is an alternative way to consider working memory and the "thought engine" of conscious mind. Perhaps everything is being handled in one giant metacircuit, which is so constructed that fragments of CIPs are sequentially accessed in conscious mind (working memory) and then used as feedstock to influence subsequent thought within that same metacircuit.

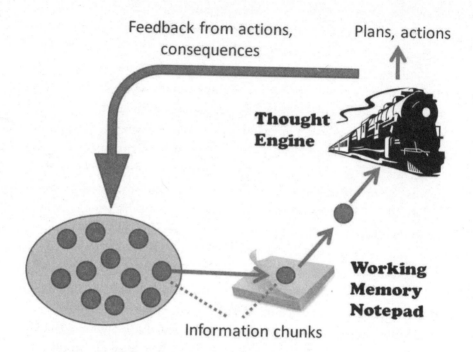

Figure 3.5. Working memory as the basis for thought; a model for how working memory participates in the conscious thinking process. Successive small chunks of information, which may come from current stimuli, memory stores, or sources generated internally from elsewhere in the brain, are successively streamed into the "working memory notepad" one chunk at a time and then into the "thought engine" circuits that accomplish the plans and actions. Think of the pad as a frame in a movie or video in which thought elements stream through the pad. Similarly, processes could also operate unconsciously.

The low information-carrying capacity of working memory is well known. This has consequences for both memory consolidation, and also for thinking ability (which correlates with working memory capacity).[19] Though formal training methods can increase working memory capacity, the conventional wisdom is that most people can only hold seven items in working memory (that is why telephone numbers are seven digits long). However, the digits in a telephone number are not strictly random and independent, and a more modern view is that most people can only hold about four random and independent items in working memory.[20]

Is there a similar *unconscious* working memory "notepad"? It does seem reasonable to suspect that the same process of feeding small chunks of information sequentially into a "thought engine" would also be an effective way for "thinking" to occur unconsciously. One possibility is that the same processes operate unconsciously, with the difference being that what is on the notepad and running through the thought engine is not accessible to consciousness. Another possibility is that there is no unconscious notepad and the flow of thought elements streams seamlessly from the information source to the thought engine. Such operation might make unconscious thought more efficient (maybe faster, but less subject to "editing").

The mental notepad, however, is crucial for conscious thought. The pad is a virtual way station where the information on it is frozen momentarily, thus facilitating reflection and decisions about how the information might be altered, vetoed, or used in alternative ways in a stream of thought. Most importantly, consciousness allows the mind to be aware of what is on the notepad and integrate it with what is about to be put on the pad. Such review and planning are hallmarks of conscious thinking.

Working memory capacity correlates with IQ. There are even studies showing that training children to increase their working memory span will actually increase their IQ. I think this is explained by the thinking model I showed in figure 3.5. The more you can hold on your virtual pad, the more information it can supply to the "thought engine." If you have more information for your brain to work on, your thinking should benefit.

It was long thought that speed of mental processing accounted for IQ, but we know now that speed makes only a relatively minor contribution to intelligence. In addition to working memory capacity, another, and related, phenomenon is that attentiveness probably increases IQ. There are multiple ways to train the brain to increase working memory capacity and to be more attentive and thus improve IQ, even in adults.[21]

One other factor seems to be the ability to identify the most relevant features in a stimulus situation. A recent study that used a simple visual discrimination task showed a strong correlation between IQ and the relative resistance to distracting stimuli.[22] Subjects were asked to distinguish direction of a single moving visual stimulus (a grating of light and dark bars of various diameters). The time that the grating was visible was adaptively varied to determine the time threshold at which movement could be detected. People with higher IQs had less ability to detect direction of motion of the larger bar patterns.

This design probably excluded any effect of attentiveness, given that there were no competing visual elements and the stimulus was on for only fractions of a second. The results were interpreted to indicate that more intelligent people can actively suppress irrelevant information, on the grounds that as the image size increased it was more likely to be registered in the brain as irrelevant background. Subjects were not asked to ignore large images. The effect occurred automatically. So, the equivalent point is that the more intelligent people have better ability to discern quickly what is important.

Small objects in the real world may be more likely to be important because they are different from the background. The authors could have bolstered their conclusion with information theory. That theory says that information content of any kind of event or stimulus is proportional to how unlikely it is to occur. In a visual field, a small object occupies only a small portion of the total field, and is thus relatively unlikely.

Two other implications need to be noted. One is that many older people are notorious for becoming more distractible (see my *Memory Power 101* book). This may reflect an age-induced decline in ability to suppress irrelevant information. One might expect a corresponding decline in IQ. Lest you want to discriminate against people because of age, let us remember that mental decline does not occur in many older people, and the majority of older people have superior capabilities because of a lifetime of experience and the learning schemas that have developed as a result.

The other implication is the issue of cause and effect. Are you less distractible because you have a high IQ or is your high IQ promoted because you have learned how to focus on relevant detail? I know of no studies in this regard, nor of anyone trying to develop a mental training program to reduce distractibility and test whether it produces IQ gains. I suspect such studies might be fruitful.

We know that intense mental activity promotes brain development (recall the section on neuroplasticity).

Consolidation of Working Memory

Working memory is a form of short-term memory that may or may not get converted ("consolidated") into longer term memory.[23] At the simplest level, working memory can be illustrated by what happens when you look up a phone number in a telephone directory. Those numbers are represented by certain CIPs, probably in multiple circuits, and the memory is accessible as long as the relevant circuits maintain that real-time representation. To do this, the circuits must reverberate to keep what's in working memory "on-line." Anything that perturbs the reverberation can disrupt the representation and thus the working memory of the numbers. That is why working memory is so vulnerable to distractions.

We know that interposing distracting or new stimuli or thoughts can interfere and disrupt an ongoing consolidation process in addition to changing what we are currently thinking about. Some people call this the *interference theory of forgetting*. The usual interpretation is that a given working memory is represented by real-time distributed CIPs, and if these reverberate long enough, the involved circuits will be facilitated long-term so that the memory engram is "laid down" in a way that enables later retrieval. Interfering stimuli and thoughts would obviously create a different CIP representation, and if this happens before consolidation, the original experience never gets facilitated and formed into long-term memory.

This consolidation process for declarative memories relies heavily on the function of the hippocampus, a C-shaped structure that is contiguous with but lies underneath the neocortex. More recent evidence indicates that consolidation also involves the medial prefrontal cortex. Most interestingly, neuronal impulse activity becomes selective during consolidation in this part of the cortex. Such CIPs are sustained during the interval between two paired stimuli but reduced during the interval during two unpaired stimuli. These new CIPs develop over several weeks after learning, even without continued training. In short, the memory is consolidated in terms of CIPs.[24]

All such CIPs and interactions may reflect a huge learning component. Firing patterns elicit other firings that have been made possible by past experiences. The CIPs of all new experiences, whether manifest in the consciousness or not, must play out in the presence of activity patterns that are elicited from stored representations. In the process, the brain can not only modify representations of what is occurring in the present but can also change the stored past representations. Each time a memory is recalled, the CIPs that represent the memory interact with the CIPs representing current experience, and the mixture is reconsolidated. This principle of reconsolidation is implemented clinically in the treatment of post-traumatic stress syndrome.

"Thoughtless" Habits and Compulsions

When thoughts are repeated again and again, they become so well-learned that they become habits. If such thoughts are associated with significant positive reinforcement each time they occur, they become so well-learned that they may lead to compulsive behavior, as in addictions of all sorts (gambling, pornography, drugs, and so on).

Bad habits and compulsions cause people to do things they would consciously prefer not to do. They may eat too much when they want to lose weight. They may sit around watching television when they know they should be out exercising. They may have bad personality traits (anger, detachment, narcissism, and so on) that they want to change but can't.

Compulsions can be overcome through force of will, but that is often insufficient. Common experience with such lack of willpower has led many people to think that only unconscious mind drives our behavior. Conscious mind, they say, cannot exert control but can only be aware of what is going on. In short, humans are considered to be like robots, where compulsions are learned and play out like computer programs. Many scholars concede that the human mind can generate intentions, choices, and decisions but nonetheless conclude that humans are not capable of free will or even of doing anything (see chapter 4).

In many human societies, alcoholism and substance abuse are major social problems, and much research has been devoted to the prevention and cure of such behavior. Many brain functions are involved, especially the brain's reward system. Substance abuse can reset this system so that the normal needs for positive reinforcement cannot be met without access to the object of addiction.

As a child, I once asked my Uncle Bob why he did not drink alcohol, because everybody else in my family drank in moderation. Uncle Bob would not drink at all, and he explained to me that he was afraid of becoming an alcoholic. Maybe there were alcoholics in his family tree. The point is that he intuitively thought his brain might too readily learn to like alcohol and he made a choice not to drink while he still could.

One recent study of cocaine addiction in rats revealed a shift from controlled drug seeking to uncontrolled seeking as the addiction progressed rapidly.[25] This shift was associated with a change in synaptic biochemistry. Rats were trained to self-administer cocaine via intravenous cannula. After a few weeks, many of those rats became addicted and compulsively self-administered cocaine. In non-addicted rats, electrical stimulation of neurons in a drug-sensitive region of the brain showed that the normal responses to stimulation changed over time; that is, these neurons had learned and "remembered" the stimulation. But in cocaine-addicted rats, this normal response change with time did not occur. The brain had been physically and chemically reprogrammed.

Let's not get into an argument over whether alcoholism and other addictions are diseases. They are diseases in the sense that the brain has become mal-adaptively reprogrammed by the object of addiction. But there was a time in the history of every person's addiction when a choice was available to preclude compulsive behavior. Even with no knowledge of neuroscience, Uncle Bob understood that.

DIFFERENT STATES OF CONSCIOUSNESS

The brain operates along an apparent continuum of states ranging from alert wakefulness to sleep to coma—the state of consciousness varies accordingly. Mental states range from the obvious (sleep vs. awake) to more nuanced states, as in sad vs. happy or hungry vs. satiated. When we go to sleep, our consciousness leaves us. We may awaken into consciousness suddenly in the morning, as if a light switch has been flipped. Dreaming, during which we are conscious actors or observers in the dream, appears abruptly several times each night, most prominently in the morning.

Switched-on states suggest a nonlinear process wherein some kind of threshold has to be reached before the state appears. What might such a threshold be based on? Perhaps it is a change in the combinatorial code of CIPs or the degree or extent of oscillatory coherence at certain frequencies. It's time for neuroscientists to examine such possibilities.

What about transitions between mental states, as when a sleeping brain wakes up? When consciousness emerges, it does so in the full flush of self-identity. When we awaken in the morning, it is then that we again realize we exist. A prerequisite for thinking, it seems to me, is for a brain to have the capacity for recognizing its embodied sense of self as distinct from its environment. By no means does this kind of self-thinking have to be a conscious endeavor. Take, for example, a simple knee-jerk reflex. This simple, two-neuron system "knows" in a most primitive sense what is "out there" and what is "in here." The first of this two-neuron chain registers sensory information from "out there" and delivers it "in here" to the appropriate muscle fibers.

What then of the "higher" levels of thought that become possible in complex circuits involving millions of neurons? Is such thought nothing more than electrochemistry and molecular biology? Now we have entered the domain of the mind-body problem. What we should focus on is the fact that the mind was genetically programmed and epigenetically sculpted to provide the circuitry necessary to distinguish self from non-self and to generate "thought" about the meaning and appropriate responses to what is "out there."

Distorted Conscious States

Hallucinatory consciousness. Conscious thought is sometimes hallucinatory. A hallucination is an unreal thought, as when we imagine hearing voices or seeing sights that are not there. Though erroneous, such thought is nonetheless consciously realized. People who hallucinate are consciously aware of such thoughts yet may not be aware of their unreality.

Is there an unconscious counterpart? Maybe, but how would we know? What we know is that hallucinations are characteristic of insanity, particularly the hearing of voices and the seeing of nonexistent images that occur in schizophrenia. One has to be schizophrenic to know what these conscious experiences are like, but we can surmise the imaginary nature from self-reports from schizophrenics.

Science cannot explain schizophrenia. Only a few clues are provided by the silent self-talk and imagined scenes that we all experience.

Normal people hallucinate when they dream. Often the dreamer knows at the time of the dream that the dream is just that, a dream, and that the dream is not real. But often we don't realize at the time that a dream is unreal. Does that mean dreaming is temporary schizophrenia? Dreaming and schizophrenia are the brain's way of staying busy inventing events and story lines. The difference is that the schizophrenic's inventions persist in wakefulness. Why don't the hallucinations in dreams persist after awakening? Maybe a treatment for schizophrenia could come from being able to answer this question.

Another important shared characteristic in normal people and schizophrenics is that both hear voices during wakefulness. Of course, the voices heard by normal people are usually their own self-talk, whereas schizophrenics hear voices other than their own. Yet normal people can hear imaginary sounds. If I go to a jazz jam session I may hear jazz in my mind's ear for hours after the session. The difference from schizophrenic imagination is that I know the sounds are memories.

Schizophrenic hallucinations are especially problematic because the patient believes the alien hallucinations are real, and the hallucinations may therefore cause the person to engage in false beliefs or inappropriate behavior. Back in the 1970s, Princeton psychologist Julian Jaynes caused quite a stir by proposing that the human brain's evolution toward the capacity for consciousness began with hallucinations.[26] According to his theory, imaginary sounds and sights began to be perceived in consciousness, and later consciousness evolved to the point where hallucinations could be seen to be unreal.

According to Jaynes, schizophrenia is the prototype of normal human consciousness, and it remains as a vestige in modern humans. He claimed that in the first human cultures, no one was considered insane because everyone was insane. While this idea seems bizarre, it does seem likely that one function of normal consciousness is to prevent and correct hallucinogenic tendencies that may be inherent in primitive brains, and, as human brains evolved to become bigger with more neocortex, conscious thinking became effective and powerful and more capable of constraining and teaching unconscious operations.

Proof for such conjecture is not possible, and Jaynes's arguments don't seem compelling to me. He even went so far as to claim that all religions began from founding prophets whose claims of hearing God or angels were hallucinations. That could be the case, but it does not support the notion that everybody in the time of the prophets hallucinated. Today most people would categorize as crazy anyone who hears God speak to them. I wonder why this would have been less suspect in the days of the prophets.

Jaynes's notion has several problems. One is the unlikely possibility that in the short span of a couple thousand years of recent history humans evolved from schizophrenic to normal beings. The genetics of whole populations don't normally evolve that quickly. Worse yet for Jaynes's argument is the fact that billions of today's evolved humans who do not hallucinate still hold religious beliefs of one sort or another. Mentally normal people still believe at least some of what their prophets may have hallucinated about.

This line of thought could lead us into the topic of the biology of beliefs, religious and otherwise, that arise as a complex consequence of experience, memory, and reason. Books on the biology of belief exist, though our understanding is incomplete and the topic is beyond the scope of this book.

Jaynes postulated that hallucinations arise in the right hemisphere and in normal humans are suppressed by the dominance of the left hemisphere (and vice versa in left-handed people). He cited a few EEG studies that showed a difference in the electrical activity in the two hemispheres, but I know of no modern studies using sophisticated quantitative EEG that address this question.

Of special interest is time-locked activity (coherence) among various neocortical regions. Since schizophrenics hear voices, hallucinate, and have disordered logic without realizing these problems, it would suggest that various parts of the brain are not coordinating well. Schizophrenic patients do have abnormal EEG coherence in resting and stimulus conditions, suggesting more diffuse, undifferentiated functional organization within hemispheres.[27]

Consciousness, as a state of mind, is not what is at issue here. People who hallucinate, whether because of brain abnormalities such as schizophrenia or because they are having normal dreams, are still consciously aware of their hallucinations. We should also bear in mind that mentally normal people can be consciously aware of hearing voices, particularly self-talk chatter, and music in their "mind's ear." The line separating normalcy and insanity may be finer than we like to think.

Dream consciousness. Dreams may be entirely hallucinatory or grounded in reality. Many people have dreams in which they are aware not only of the dream experience but also that the events are part of the dream and therefore not real. Dreams have to be a special form of consciousness, maybe not too different from ordinary consciousness. In both dreaming and wakefulness, the brain is activated, as indicated by EEG patterns.

Common experience teaches that dreams are a special form of consciousness. Though we are behaviorally asleep, the dream content is a conscious experience, though we may not remember it after awakening. Such forgetting is a memory-consolidation problem, not a consciousness issue. We don't remember the dream because its memory was erased as we fell back into regular sleep or were flooded with new distracting thought as we awakened.

Another reason for linking dreams with the consciousness of wakefulness is that both states are egocentric. Dreams almost always involve a person's sense of self, either as a participant or as a witness in a simulated world. This is just like the sense of self's engagement with the real world. Dreams thus are a brain's way to nurture its ego, to continually "remind itself" just who the brain has created. Likewise, one could think of dreaming as the brain's way to practice its habit of generating and sustaining the conscious sense of self. This is likely part of the development process in the late-stage fetus. Also, during dreaming the brain is adding to its storehouse of information about its self. That sense is so important to individual survival that the brain works day and night to nurture it.

Let us consider animal dreaming. Anybody who has ever watched a sleeping dog bark and paddle its feet while sleeping can have little doubt that they are chasing a critter in their dreams. Sleeping dogs will even sometimes twitch their nose, suggesting that they even have olfactory hallucinations. All higher mammals, and to a lesser extent birds and higher reptiles, show multiple episodes during their sleep where bodily signs are identical to those of human dreaming:

an EEG showing low-voltage, high-frequency activity; rapid eye movements; irregular heart and respiratory rates; and spastic twitches of muscle.

Are higher animals thinking consciously when they are awake? No one can know (except the animals), but there are many books that argue both sides of the possibility. Given that animals have less developed brains than humans, according to Jaynes's view, we might think that the waking state of higher animals would be perpetual hallucination. Their dreaming indicates that their brains have the capacity for nonlinguistic hallucination, but that is not proof that hallucination is the default mode of operation in the waking state. Since higher animals, especially performance-trained animals, can exhibit a great deal of adaptive, purposive waking behavior, it would suggest that they are not hallucinating.

Where Mind Goes When We Sleep

When I go to sleep at night, I wonder, "Where did I go?" Except for my occasional re-emergence in dreams, my "I" has vanished. Yet I wake up in the morning with my "I" intact, except for the dimly perceptible changes in memory consolidation that occurred during sleep. Why don't I wake up as somebody I would rather be (as a teenager I would rather have been Stan Musial). No, I am always stuck with my usual "I" because that is the representation my brain has constructed in long-term memory over the years. There are only small, incremental changes as I age, which might conclude someday in the terrifying possibility of Alzheimer's disease.

When the "self" goes away when you fall asleep or are anesthetized, it disappears because the CIPs that sustain it are temporarily inactive. Why and how can the self come back when you wake up? It comes back because the CIP for self is actually a memory that can be recalled.

Where does this leave the relationship between conscious and unconscious mind? When normal humans are awake, they typically are automatically conscious. Given that unconscious and conscious functions are so different, I submit that each must have different CIPs, which should in theory be identifiable. Alternatively, the CIPs may be the same in both states, but what differs is the degree of synchrony of activity within circuits.

One good way to study the relationship of conscious and unconscious mind is to study the unconscious state of sleep and the transitions of consciousness to and from sleep. It is no accident that real sleep only occurs in advanced brains. Several decades ago I surveyed research reports on brain-wave (EEG) signs of sleep in various species. Brain indicators of sleep do not occur in fish, amphibians, or primitive reptiles. Only in advanced reptiles, birds, and mammals do physiological signs of sleep occur, and full-blown sleep only occurs in advanced mammals. I discuss these studies in another context in chapter 4.

Scientists learned decades ago that neurons do not shut down when you go to sleep. Many neurons fire impulses just as actively during sleep as they do during consciousness, and some neurons are even more active during sleep. What most likely changes is interval patterning among impulses and the degree of interaction and communication among neurons. As humans go to sleep the communication among cortical neurons breaks down.

Communication breakdown during sleep has been demonstrated experimentally.[28] The researchers recorded EEGs from many scalp sites over the cortex while at the same time stimulating a small patch of right frontal cortex with transcranial magnetic pulses. The responses to this stimulation at various other cortical locations indicate how well information is being communicated throughout the cortex. They found that, when awake, the stimulus triggered responses in sites nearby the stimulation site and also in similar structures on the opposite side of the brain. However, during sleep, the stimulation only evoked responses at the stimulation site. In other words, during sleep, brain areas seem to stop talking to each other. Consciousness likely requires such communication. The investigators have not yet performed the comparable experiments during the dream stages of sleep, when a kind of consciousness is present. My bet is that wider communication between areas is restored during dreaming.

Another unstudied, but probably important, dimension to this matter is the degree of timing coherence of brain activity in different parts of the brain. I suspect that activity in specific frequencies is much more phase locked and coordinated at multiple cortical sites during wakefulness and dreaming than during nondream phases of sleep.

I will expand on the notion that the sleeping brain is active, just in a different way. Whatever thinking we do during sleep is unconscious, such as processing the limited sensory input the sleeping brain receives (sounds, touch sensations from bedding, and so on). We know the brain thinks during nondream sleep because certain memories are consolidated then.

It is true that the sleeping brain is not processing much new information because it isn't getting much new information. There is also no conscious mind accessible to the sleeping brain to help inform the brain. This suggests a function for conscious mind. Namely, it is a primary source of instruction and programming for unconscious mind.

Sleep shows us that unconscious mind operates in two ways. It operates one way during sleep, when the unconscious mind operates in stand-alone mode, and another way during wakefulness, when consciousness is present as a concurrent state wherein the "two minds" can interact.

Many scholars think there is one-way interaction between the unconscious and conscious minds. That is, the unconscious mind informs consciousness of some of what the unconscious mind is thinking (see following section on free will). But I think there is a two-way alternative in which conscious mind provides a thinking resource not available in unconscious mind and thus is a major source of programming and "enlightenment" for the unconscious mind.

Our thinking about sleep and wakefulness is being pushed in new directions by studies in which human epileptics with electrodes implanted in their brain have their brain electrical activity during sleep compared to their brain electrical activity during wakefulness. Of particular value are the studies that simultaneously monitored EEG signals as well as nerve impulse activity.[29] These studies reveal the prominence of large slow-voltage waves during nondreaming sleep (slow-wave sleep, SWS), and they confirm Michael Steriade's original observations that there are also small, high-frequency gamma waves riding on top of the large slow waves. What is new are the observations of several studies showing that in SWS both impulse activity and gamma waves are phase-locked to the slow waves. It is not clear whether the phase locking of impulses to slow waves occurs because the impulses are driving the slow waves or because slow waves impose propagation constraints on the impulse traffic that gives rise to gamma waves. (Note that impulses provide much of the current underlying field potentials and in turn are influenced by them). Note also that some field potential current comes from glial cells.

What do the observations tell us about sleep and wakefulness? It suggests that gamma-generating processes and slow-wave generating processes may be mutually antagonistic. Slow-wave processes of sleep dominate electrically because they are so large. It may well be that the concurrent gamma processes enable memory consolidation in spite of the presence of slow waves. In fact, slow-wave processes may serve a useful function if they prevent the distractions and memory-consolidating interferences that are inevitable during wakefulness.

Another way to think about the gamma activity is that it may be constantly pushing the brain to wake up, but it may be blocked from doing so in the presence of SWS processes. In other words, consciousness can't occur when gamma activity is impaired and impulse throughput is diminished.

The brain insists on sleeping, and this suggests that slow-wave processes presumably satisfy whatever it is the sleeping brain needs. The traditional explanation is that we sleep to rest the brain. But nobody has documented what kind of rest the brain is getting during sleep or how predominance of slow waves could provide it.

Before the discovery of concurrent gamma activity during SWS, many scientists assumed that neurons were less active during sleep. Maybe it is only a few select neurons that need periodic rest. Neurons that fire less, for whatever reason, make fewer metabolic demands and are given more time to regenerate their energy reserves, to actively transport ions against gradients, and to rejuvenate synaptic metabolic processes. Maybe it is the glial cells that need metabolic "rest."

What sort of "rest" or benefit comes from phase locking of slow-wave generating processes? Nobody knows and, of course, nobody asked until just recently when the phenomenon was discovered. We can speculate that phase-locking is a pacing mechanism for the underlying neuronal processes. Maybe the phase-locking itself is a rest process.

At another level of analysis of the rest effect of sleep, very recent research indicates that channels in the brain open to allow up to a 60 percent increase in circulation of cerebrospinal fluid into the interior zones of the cortex.[30] Thus, sleep provides a way to dissipate accumulated toxic compounds that build up during wakefulness. These include beta amyloid, adenosine, and glutamate neurotransmitter. During wakefulness, the release of norepinephrine transmitter in the cortex may keep these channels closed.

Sleep also does something else. It promotes the expression of several genes that are involved in synthesis of cell membranes, particularly those in the oligodendrocytes, the brain's support cells.[31] Unlike neurons, oligos die and are continually replaced in the brain. Thus, anything that affects their turnover is important for brain function. Sleep has been implicated in this turnover because a common neurotransmitter in the brain, glutamate, is known to increase in wakefulness and decline during sleep. Glutamate suppresses maturation of oligo precursor cells into formation of myelin insulation. And myelin is essential for normal propagation of impulses through circuitry. To illustrate, multiple sclerosis is a disease caused by damaged myelin.

Memorizing in Our Sleep

We do know that thinking and use of working memory occur during sleep. Sleep also helps to consolidate the short-term memory of a day's events into long-lasting form.[32] Abundant research in the last few years has shown that memories of a day's events are being processed when you sleep; that is, when only your unconscious mind is operative. While conscious mind sleeps (gets a rest?), the unconscious mind stays on the job. During sleep, our unconscious mind is allowed to work without all the interferences that our conscious mind generates during the day.

In one study, students were paid to stay awake all of one night and then try to learn thirty words the next day.[33] The students were subsequently allowed to catch up on sleep loss and were tested on the words. Compared to a control group that was not sleep deprived prior to the learning session, sleep-deprived students remembered 40 percent fewer words. Sleep-deprived students also tended to remember positively connoted words far less accurately than negatively connoted words. This study implies the presence of "proactive interference," wherein sleep loss before learning a task affects memory consolidation. Additional studies, such as one involving image recall, support the idea of "proactive interference," and they have even shown that brain areas involved in memory, such as the hippocampus, are more active in subjects who get a normal night's sleep.

Another recent sleep study has linked sleep with superior consolidation of motor skills.[34] Using the premise that memory of motor learning develops without practice ("off-line") after a learning session, this study taught two groups of subjects a simple motor skill; one group was taught during the morning and the other during the evening. A learning disruption in the form of magnetic stimulation to the motor cortex was delivered to each group of subjects, yielding results indicating that the memory interference could not be compensated for by off-line learning during the daytime but was compensated for in the night-time group. This observation leads to the assumption that during the daytime, numerous memory-disruptive influences interfere with off-line consolidation of material that we learn in the morning, but with night-time learning there are far fewer disruptive sensory and cognitive influences because we are asleep.

The advantages offered by having fewer disruptive influences during sleep have also been confirmed in a brain imaging study.[35] Sound stimuli were presented to sleep-deprived patients during nondreaming sleep. The results indicated a suppression of activity in the auditory pathways that the researchers were attempting to stimulate. However, results also indicated suppression of activity in the visual cortex, suggesting that sleep protects the brain from the arousing effects of external stimulation during sleep, not only the primary targeted sensory cortex but also other brain regions that are interconnected with visual cortex. It is in blocking out such interference effects that sleep helps facilitate consolidation. This study also prompted researchers to conclude that consolidation of memory occurs over many hours (at least in sleep-deprived subjects) rather than over the course of just a few hours.

Activated Sleep

In 1953, University of Chicago researcher Nathaniel Kleitman and his students Eugene Aserinsky and William Dement monitored people as they slept and reported that sleeping humans go through periods during which they exhibit profound reduction of muscle tone with superimposed twitching of limbs, erratic breathing and pulse, rapid eye movements (REM), and EEG patterns like those seen when a person is awake. This obviously was a different kind of sleep, and it came to be called activated or paradoxical sleep. Later studies confirmed that if you awaken sleeping humans during REM they invariably will report that they had just been dreaming.

Physiologically, we know that dreaming is the hallmark of advanced animal evolution. It is fully developed only in mammals, whose brains have a well-developed neocortex capable of "higher thought." Paradoxically, babies spend more of their sleep time in dreams than do adults, yet their neocortex and fiber-tract connections are poorly developed compared to adults.

Why do we dream? Books have been written on the subject, but we still don't know. We do know that the stage of sleep in which dreams occur is a necessity. Many animal and human experiments show that the brain does not function normally if it is not allowed to dream. A whole array of reasons for dreaming have been suggested, and they are not necessarily mutually exclusive: (1) to ensure psychic stability, (2) to perform off-line memory consolidation of events of the preceding day, and (3) to restore the balance of neurotransmitters that has been disrupted by ordinary nondream sleep. Dreaming may also just be an inevitable side effect of the reorganization of unconscious mental processes.

Maybe, like our dogs and cats, human dreams just inevitably accompany a physiological state that is just the brain's way of stimulating or entertaining itself. In any case, dreams can be very good indicators of what is on our minds, though the pronounced symbolism in dreams may require a good deal of introspection and analysis to interpret. Why are dreams so often symbolic rather than literal? Nobody knows, but maybe symbolism is a result of unconscious thinking trying to become manifest in the special consciousness of the dream state.

All human brains dream. Those few people who claim not to dream are wrong. They dream; they just don't remember dreaming, as has been documented in experiments where people are awakened immediately after their brain and body signs indicate they were dreaming. Failure to remember is probably explained by consolidation interference from a flood of distracting sensations at the moment of awakening.

After the discovery of these dream-sleep signs in humans, Dement observed that these signs occurred periodically during the sleep of cats. Anyone can see the REM and muscle twitches in sleeping cats and dogs, but it wasn't until Dement recorded brain waves that we knew for certain that animals, at least higher animals, must also dream. Of course, you can't prove that dogs and cats dream, but it is a reasonable conjecture.

"Higher" animals (all mammals) show all the same outward signs of dreaming as we do. My student, Al Prudom, and I even demonstrated this in one of the most primitive extant mammals, the armadillo. Periodically during their sleep, the EEG becomes "activated," the eyeballs flit around erratically, and muscles twitch.[36]

I suspect that advanced mammals like dogs and cats are conscious observers or participants in their dreams. Consider the following scenario: You finished dinner, had dessert, and fed the dog. As you flop into the soft recliner to relax and watch a little TV, you notice your dog, having no entertainment outlet, take her customary nap. Before too many commercials pass, you hear claw scratching on the floor and muffled barking bordering on whimpering. The dog is also padding her feet, as if trying to run. If you look closely, you will see darting of her eyes. You surmise she is dreaming, and if you had electrodes on her head to record her brain waves you would know for sure, because the brain would be dominated by high-frequency brain waves.

You ask yourself, *I wonder what she is dreaming about? I bet she is chasing that pesky squirrel that drives her crazy every day.* By day, the squirrel always escapes. By night, your dog gets to try again. *Why do dogs dream?* you wonder. *Well, it is entertainment*, you think, adding, *that dog is so lazy, I bet she sleeps because it is an easy way to entertain herself. She has learned that going to sleep always leads to some unexpected adventure.*

We know that REM sleep is required by the brains of higher animals and humans. Forced deprivation of dreaming causes irritability, emotional upset, and dysfunctional thinking and behavior. In people, dream deprivation can drive one to the edge of insanity. If you deprive people of REM for several nights by awakening them each time they exhibit REM signs, they will have an abnormal incidence of REM on subsequent nights when you leave them alone. It is as if loss of REM creates a debt that has to be repaid.

One more recently discovered fact about REM sleep is that memories are being formed during this stage of sleep. Many studies have shown that deliberate disruption of REM sleep impairs the formation of certain permanent memories of the events of the day before the dreams were disrupted. Memories are formed by repeated firing along specific distributed circuits in the brain. Perhaps these firing patterns are regenerated, augmented, and sustained during dreaming.

Dream content is likely a byproduct of that activity, not its cause. This does not mean that what we dream about is what our brain is trying to remember. Common experience teaches that many of our dreams have no apparent relationship to what happened the previous day. How can we explain this paradox? We can't. However, dreams would not necessarily have to reflect the events being remembered. For example, perhaps activity in memory circuit *A* triggers activity in another circuit, *B*, which is involved in generating other dream content.

Some dreams may be a way for the unconscious mind to grab the attention of the conscious mind. There are many things that the conscious mind does not want to think about, which no doubt includes things that the unconscious mind is trying to raise to the surface. Even though dreams are often highly symbolic, they provide a way for the unconscious mind to make thought available to the conscious mind for more rigorous analysis and interpretation. This is also another way of stating the understanding about dreaming that Freud gave to the world. So we should pay attention to our dream content, though doing so is not easy. And while numerous books have been written to help us interpret dreams, for our purposes here, it may suffice just to point out that the unconscious mind is very real, and that it is in constant interaction with the conscious mind, especially during dreaming.

I haven't yet answered the question of why we dream or, better stated, why our brain needs to generate episodes of the REM stage of sleep that enables dreaming. I phrase it that way because I don't think we dream in order to generate activated sleep; rather, I think that because the brain has activated sleep, dreaming is easy to generate. There must be some reason for dream sleep, but scientists don't know for certain what it is. Speculation abounds. Maybe this activated form of sleep is needed to restore chemical or circuitry balance that has been in low gear all during the regular sleep periods. Maybe activated sleep, which follows the restful sleep where one "falls into the pit," may be a way of protecting the brain from falling too far into oblivion, or maybe activated sleep is a form of rehearsal to get ready for the next day of wakefulness (note: people who sleep at night dream the most in the early morning).

But why do only humans and certain other animals experience activated sleep? And why so much—about two hours' worth every night for humans, broken up into multiple segments interrupted by regular sleep? There must be a more fundamental explanation for us and all the other animals that have the ability to dream.

Here are three common theories for why we dream: (1) to help normalize stress and emotions, (2) to consolidate memories, (3) to get ready for the next day of consciousness. The leading theory is that dreaming has no purpose; rather, it is just random neural activity that causes distorted thinking that manifests as fabricated stories.

Historically, all dream theories have arisen from the perspective of the dream interpreter. Ancient shaman priests and religious prophets have thought of dreams as God's way of communicating with us. Of course these folks didn't know that many animals also dream. If they had, they would have had to conclude that God talks to animals, too. Any good theory for dreaming must accommodate two basic facts: (1) signs of dreaming occur only in advanced species (more recently evolved reptiles and mammals, with the most robust dreaming occurring in humans), and (2) in every mammalian species dreaming incidence is far more frequent in babies than in adults, and dreaming also decreases with age.

Among the early ideas that have been mentioned is the possibility that dreaming acts as an escape valve for letting off the "psychic steam" that accumulates from emotional stress during the day. Psychiatrists think of dreaming as the brain's release of unconscious thinking into dream consciousness. Thus dream analysis can be a key part of psychotherapy.

Dream content does have symbolic meaning, as Freud showed, but that is a consequence of dreaming, not its cause. Also, modern studies of young people show that about seven out of eight dreams are unpleasant, with about 40 percent of them being downright frightening.

Another theory, currently in vogue, is that dreaming is needed to help consolidate memories of the day's experiences. Experimental disruption of dreams does interfere with long-term memory formation. However, memory consolidation also occurs during deep stages of nondream sleep.

Today a popular explanation for dreaming is the "activation-synthesis theory," which posits that dreaming occurs because cortical neurons fire randomly, causing an attempt to construct a story line that makes sense of the random impulse firing patterns of sleep.[37] However, I am not aware of any statistical tests that confirm random firing during dreaming, and, moreover, many neurons have distinct firing patterns during dreaming and are therefore nor firing randomly.

The story line notion, however, is relevant. Human wakefulness consists of episodic experiences, which to the brain probably seem like a series of adventures and stories. The dreaming brain probably does try to reconstruct the processes it has learned during wakefulness that characterize what conscious life is all about. The result is abortive dream stories that simulate real conscious life. The stories are often incomplete and irrational because they are abortive attempts at consciousness, and they lack the corrective "reality checks" that occur with wakeful experiences. Notably, dreams engage our sense of self, either as an observer or as a participant. Dreams bring the sense of self into being after its banishment in deep sleep.

All theorizing of this kind misses the obvious point about why people dream. The answer as to why we dream is simple. We dream because we are in activated sleep, which promotes the nonlinear dynamical processes that support the conscious sense of self. We don't dream to vent psychic steam; we vent psychic steam because an activated brain can provide conscious access to submerged emotions. We don't dream to consolidate memories; we consolidate memories because that is promoted by being in an activated brain state. We don't dream because the brain wants to create explicit story lines; it creates story lines because that is what an activated brain can do.

Early in my research career as a physiologist I codiscovered dream sleep in ruminants, which at the time were regarded as capable only of dozing and cud chewing. Notably, I showed that they not only demonstrate all the typical sleep stages, but also that their rumen activity stops during sleep, lest they drown on their own cud.[38] The important point was that if an animal has genuine physiological sleep, it must also have REM episodes. I concluded that dream sleep was the brain's way of getting itself ready for shifting from sleep to the wakefulness needed for the forthcoming day's activity.

I thus saw dreaming as a "dress rehearsal" for the next day, which is an integral part of my idea that the brainstem enables a brainstem-mediated "readiness response" to stimuli to assure that appropriate behavior and cognition occur.[39]

Back then I looked at the issue from the perspective of an animal physiologist. A similar explanation for dreaming was proposed in that era by Bob Vertes.[40] Each of us is a physiologically oriented neuroscientist, and in fact we have collaborated in editing the book referenced earlier on brainstem function.

The "dress rehearsal" idea still appeals to me. But it does not completely specify why the dress rehearsal is generated. In more recent years I have developed an interest in consciousness and cognition. That orientation leads me to a new way of looking at dreaming, one that conveniently is compatible with the initial view that activated sleep helps get us ready for a day of conscious activity. It is also compatible with other theories for dreaming. And best of all, it fits with my idea of how the brain creates consciousness.

Historically, the problem has been trying to explain a purpose for dreaming. But the focus on dreaming has things backward. We need to focus on why we have activated sleep, which produces dreams as a side effect. Activated sleep could help the brain "remember" and rehearse how to reactivate readiness when the time comes in the morning. A related explanation for dreaming in young, immature brains is that the brain activation provides a source of endogenous stimulation that is useful in promoting brain maturation and the capability for consciousness. Realize that baby brains have not yet learned how to wake themselves up in the absence of stimulation. Their brain is likely inefficient at waking itself up.

* * * * *

In the early morning of January 22, 2010, I stumbled upon the key to an explanation. I did it in my sleep. More specifically, a new idea came to me as I awoke from one of my own dreams. I asked myself, "Why is it that in the morning, I awaken while dreaming?" When I checked with other people, they too reported the same morning experience. This caused me to think about what happens when a human goes to sleep. According to research in the animals I have studied and the reports of human studies, early in the night the brain tumbles into a very deep sleep, a deep abyss of sleep that is as far removed from consciousness as a normal, undrugged brain can get.

If you have ever been to a sleep lab to check for sleep apnea, as I have, the technician will tell you that in the first hour after going to sleep, your brain shuts down so profoundly that you may even stop breathing. You would actually die (and a significant number of people have), if there weren't reflex mechanisms in the brainstem's "nonconscious brain" that force you to breathe when blood oxygen gets dangerously low.

This deepest stage of sleep of is called slow-wave sleep (SWS) because EEG waves are of relatively long duration). SWS occupies about half of the first hour of sleep, and it has a high incidence during the next two hours. Most of the first three hours are stages of mental oblivion.[41]

Whatever brain circuitry and patterns of nerve impulses that were needed to construct and sustain consciousness during the daytime are obliterated when one falls into that SWS pit. To wake up without any outside stimulus, the brain has to figure out how to get back what it lost; that is, the really complicated task of generating conscious mind. How does the brain get itself out of this abyss? It's a steep hill to climb. No meaningful stimuli are triggering the brainstem arousal system into action, and a sleeping brain has no easy way to activate the brainstem's arousal system to wake itself up. External stimulation is normally required to do that. Typically, we try to avoid external stimulation so we can go to sleep and stay there.

If there is no stimulus to cause awakening, such as an alarm clock (or the need to urinate), the problem for the brain is how to recover from SWS's demolition of all the processes it used to create consciousness. It is as if consciousness is a Humpty Dumpty sitting on the highest wall of brain function, and SWS shoved it off and smashed it. Who can put it together again? "All the king's horses and all the king's men" can't do it. Incidentally, I first presented this "wake-up" Humpty Dumpty (HD) theory at the annual meeting of the Society of Neuroscience, where many foreign scientists didn't get the Humpty Dumpty metaphor because that fairy tale was not part of their culture. The scientific explanation and defense has been published in a peer-reviewed journal.[42]

To put this wake-up theory another way, the sleeping brain, as with a computer, has to be booted up, and activated sleep is a crucial part of the process. Unlike booting up a computer, slow as it may be, rebooting the brain takes all night. It is as if dream and nondream sleep are in competition, and as a night's sleep progresses, dream sleep progressively wins out. This maxim seems to be confirmed by common experience. Ever toss and turn at night because your conscious mind is too busy thinking of ideas, problems, or strong emotions? Well, dreaming produces a similar effect during activated sleep episodes. Dream content is consciously perceived, and this generates robust thinking processes that compete with and eventually overwhelm the fog of unconsciousness.

This is not to say that the brain says to itself, "I am in deep sleep, I must find a way to wake myself up." Such a teleological explanation is not necessary. It may be that the default mode of brain operation is wakefulness. After all, it is wakefulness that assures a human will eat, drink, and do the necessary things to stay alive. In this view, sleep may be something the brain is required to experience periodically to sustain the capacity for normal wakefulness. The common assumption is that the brain needs rest, but what that rest entails is not at all understood. It is clear that many neurons are active during SWS. It is likely that SWS enables clearance of toxic metabolites accumulated during wakefulness. We don't have to understand what "brain rest" is to speculate that dreaming may just be the brain's way of breaking the constraints of sleep and "rebooting" to get back to its default mode of operation.

It is no trivial matter for a comatose brain to recover from a deep SWS episode that demolishes the required processes for consciousness. In the absence of external stimulation, putting "Humpty Dumpty back together again" must be quite a challenge. The progressive increase in dream sleep as the night wears on may serve to help the brain remember and rehearse how to implement the processes needed to lift the brain out of the pit and get it ready for the coming day's conscious activities.

When activated sleep and dreaming are triggered, the brain is trying to reconstruct the normal CIPs of self-aware consciousness. Because it has not yet succeeded, bizarre dreams and aborted attempts at awakening result. Maybe it hasn't succeeded because the toxic metabolites have not yet been cleared. Rather than being dominated by random impulse activity, the brain is methodically trying to reconstruct the combinatorial coding and coherences of neocortical circuitry needed to facilitate awakening into full conscious competency. This reconstruction probably includes adjusting the sleep-wakefulness servo-system's set-point circuit dynamics to make consciousness easier to generate and sustain, so as to be fortified against drowsing off to sleep during boring parts of the day.

Here I am focusing on what must happen in sleep, during which consciousness disappears because there is no excitation coming from the outside world. The problem during deep SWS is compounded by the loss of excitatory drive to the brainstem arousal system. In sleep, this arousal system doesn't get much stimulus. The only way it can be turned on is from feedback from the neocortex.

The neocortex and the brainstem arousal systems do have a reciprocal relationship. That is a reason why it is so hard to calm down after a highly stressful or emergency situation. The brainstem triggers cortical activation, and an activated cortex supplies feedback to help keep the arousal system active. In the waking state, these effects are augmented by the steady stream of sensory input to the brainstem reticular formation that occurs during wakefulness.

But in deep SWS neocortex neurons that supply feedback input to the brainstem slow their impulse firing drastically, thus removing excitatory drive to the brainstem, which in addition is not getting much stimulus from outside the body. If a sleeping brain is to become conscious without external stimuli, it must have a self-generating mechanism, and that mechanism probably requires the cortex to activate the brainstem—just the reverse of what happens during wakefulness.

The projections from the cortex to the brainstem have recently been demonstrated in studies in which EEG and fMRI scans were taken simultaneously in humans as they performed auditory and visual tasks. Strong reciprocal connections were observed between the cortex and the brainstem.[43]

I submit that the mechanism for activating the brainstem and in turn the neocortex during sleep is provided by the mechanisms that trigger activated sleep. That mechanism must include the capacity to increase activity in certain neuron clusters in the pons part of the brainstem. These neurons produce the signs of dreaming, such as rapid eye movements and intermittent suppression of bodily movement, and they probably also cooperate with adjacent brainstem neurons of the arousal system, which in turn launches activated sleep and the consciousness of dreams. These neurons increase their firing during dreaming and cease it with the transition to wakefulness. Thus, the main difference between wakefulness and dreaming is that these special brainstem neurons are only active during dreaming, while most of the other reticular-formation neurons are active during both dreaming and wakefulness. The main difference in the nature of the consciousness of the two states may be that in wakefulness the brain has access and can respond to real-world stimuli. In any case, it would seem easier for the brain to switch from dream consciousness to wakefulness than from SWS to wakefulness.

Brains have to know how to generate the right impulse patterns and timing relationships throughout vast neural networks involving hundreds of millions of neurons. They have to know what oscillatory frequencies to use and where, and how to synchronize them. In deep SWS all this has to be done from a zero baseline. It takes an entire night of repeated dreaming for the brain to do this on its own if it is unperturbed by external stimuli.

To have a way to wake itself up, the process apparently has to occur in stages, moving from SWS to lighter stages of sleep, and then to activated sleep, which at first occurs in short abortive episodes. As the night progresses, the brain gradually emerges from SWS into lighter stages of unconsciousness, and then the sleep is periodically interrupted by activated episodes that become more frequent and longer in the early morning.

One reason SWS may be hard to terminate is because it is not finished with whatever essential tasks occur during SWS, such as regenerative biochemical "rest" for neurotransmitter systems. Another reason a sleeping brain has to struggle to wake up could be that it is busy working on forming certain memories, and it has to do this under the handicap of minimum functional capacity. Neural resources for self-awakening may not yet be available.

This brings us to the issue of learning. Is it not possible that a brain has to learn how to arouse itself from sleep? Think about babies. They sleep most of the time and they spend far more time in dreaming than adults do. About 50 percent of their sleeping is spent with the physiological signs of activated sleep, compared to only about 20 percent for adults. Yet babies have little to dream about and few life experiences that would require the ventilation of psychological steam in dreams or the performance of much memory consolidation. For babies, the brain is still getting used to the experience of consciousness. Generating consciousness can be thought of as a skill that the immature brain barely knows how to accomplish. After all, because of physical limitations, a baby has very limited opportunity to hone its consciousness-generating skills through conscious interactions in the outside world. The inner-world conscious engagement provided by dreaming may be nature's way for the brain to master one of its most important tasks: triggering consciousness in the absence of external stimulation.

Adults dream less frequently as they get older. Supposedly, the adult brain has become pretty adept at generating consciousness. In the process of spontaneous awakening, the brain must consult its memory stores to revisit what has been acquired through years of learning how to expedite the consciousness reboot. Thus, it shouldn't take as much activated sleep for adults to reach the threshold for wakefulness.

In the SWS pit, the representation of self is abolished. It has to be reconstructed from memory. Activated sleep could be the brain's way of reinventing the consciousness wheel each day. In generating activated sleep, the brain is grabbing bits and pieces of the consciousness-generating process, producing a dream consciousness that is incomplete and not fully controlled. Nonetheless, these early attempts may help the brain recover the capacity for creating consciousness. These attempts may have to reach a certain threshold, and may even be probabilistic, involving trial and error. Thus it might take a series of activation episodes throughout the night for the brain to put Humpty Dumpty together again.

A summarizing metaphor. One way to think about all this is to think of being awake as being in a brightly lit room. When you want to go to sleep, you switch off the light switch and become enmeshed in pitch blackness. Sleep is another way the brain finds that "it's dark in here." Now, after a while you want to wake up, but you are disoriented and don't know where the light switch is. So you light a birthday candle, but it does not produce enough light to see the switch. You have learned even when you were a child that if you light enough candles you can find the wall switch. So you anchor the candle to keep burning and move to another part of the dark room and light a candle there. That didn't help enough either, so you anchor that candle and move to another area. Eventually, you have enough candles lit that you can see where the wall switch is. The rest just requires a flick of the wrist.

Fitting known phenomena into the new explanation. How do we reconcile this new view with the obvious fact that arousing stimuli during sleep, as with an alarm clock, can wake you up without the need for preceding dreams? First, if you are in SWS when the alarm goes off, you are not easily aroused because the depth of sleep in that stage is quite profound. Second, even when awakened, most people are still groggy and not at their peak level of conscious function.

External stimuli awaken us because their collateral inputs into the brainstem reticular formation induce a cascade of arousing influences that spread into the reticular nucleus of the thalamus, and from there into virtually all regions of the cortex. Lacking such a mechanism in sleep, internally generated awakening must necessarily involve tentative fits and starts that find expression via dream sleep.

In spontaneous awakening from sleep, the brain has to find a way to engage these arousal processes, which may not be all that trivial in the absence of external stimuli. It's not the dream content that wakes us up. It is the reticular formation activation that generates the reboot; the dreams are incidental.

Can the Humpty Dumpty theory explain why we resume a normal dream-sleep pattern after being awakened in the middle of the night? When you are awakened in the middle of the night, for whatever reason, your brain still needs more sleep, which is why you can get back to sleep.

Activated sleep may also adjust the sleep-wakefulness servo-system's "set point" of brain circuit dynamics so that consciousness is easier to achieve and sustain, being fortified against drowsing off to sleep during boring parts of the day. Part of this preparation may well be a final retouch on consolidated memories to store them in a way that optimizes recall access by conscious working memory.

How do we explain why only mammals and a few advanced reptiles engage in activated sleep? Here, the explanation is that primitive species lack the neocortex piece of the neural machinery to produce robust consciousness in the first place. Thus, there is less need to have a "rehearsal" process for the next day's activities. Moreover, brainwave studies suggest that the brains of these animals don't have the mechanisms for the slow-wave kind of sleep either; that is, they don't fall into a deep-coma pit from which their brains have to climb out.

SWS and dream sleep apparently coevolved, reaching an apex in humans. SWS may be a needed recovery process in species that can generate full consciousness, and activated sleep episodes could be part of a needed recovery process for SWS.

What does incomplete activated sleep in lower mammals indicate? Their activated sleep episodes are short and infrequent during a night's sleep. Because their wakeful consciousness is not as developed, their brains do not have as far to go to restore wakefulness from SWS. But the fact that they have active sleep at all suggests that they have some degree of consciousness during wakefulness.

They may be sentient beings, though at a lower level than humans. Anybody who has had close relationships with animals such as dogs, cats, or horses already appreciates this possibility.

How do we explain why the most developed activated sleep capability occurs in humans? A species that has the highest level of consciousness has more need for deep sleep and has farther for its brain to go to lift brain function out of the SWS pit. It may also be true, though it has never been tested to my knowledge, that SWS is deeper and more profound in humans than in lower mammals. More activated sleep may be required because it is harder to generate a higher level of consciousness.

How do we explain research results that show that memory consolidation occurs both in slow-wave and in dream sleep? One possibility is that the nature of the memory determines whether consolidation is augmented preferentially during SWS or during dream sleep. Certain kinds of memories may be preferentially fortified during the respective stages of sleep. Presumably, the memories that benefit from SWS are more relevant to the explicit, episodic processes that are expressed in the context of the human ego and sense of self.

This possibility may actually be testable. Would there be selective memory disruptions if experimenters selectively deprived people of SWS? But depriving people of SWS is hard to do, since brains fight to get their SWS experience. I would also predict that under such conditions there would be less dreaming (and more tired and irritable consciousness). But it is easy to disrupt dream sleep selectively, and it is thus possible to structure experiments to find out just which kinds of memories are less able to be consolidated as a result.

Why doesn't the brain try to recover from SWS with just one long activated sleep episode instead of a series of choppy short episodes? I suggest that activated episodes cannot be sustained early in the night because the need for SWS has still not been met. The repeated short activated episodes probably reflect the fact that it is not easy for the brain to fight its way out of its stupor.

Why do you and I wake up right at the end of a dream? Review of my own dream content over many decades of dreaming convinces me that it is not any exciting, frightening, or other attention-grabbing aspects of dreaming that wakes me up. On many occasions, I have awakened from boring and insignificant dreams when they occurred during my normal wake-up time. More likely, we wake up because the brain has achieved the final threshold for completing construction of consciousness. At this stage of a night's sleep, all the brain needs to do is lift the remaining obstacles to receiving external input and switch off the few neurons that are controlling the unique features of activated sleep, such as eye movements and movement suppression.

Two phenomena are not so easy to explain in "Humpty Dumpty" terms: narcolepsy and awakening from anesthesia. But both can be accommodated. First, it should be noted that neither phenomenon is representative of normal brain function. It may be too much to expect any theory of normal function to explain all abnormal functioning.

In the case of anesthesia, the problem is explaining how we seem to emerge from anesthesia so suddenly. In my experience of recovery from anesthesia, consciousness has just "popped up" with no previous dreaming (that I know of). First, the impression that consciousness just pops up may be an illusion. The anesthetized brain must struggle mightily to overcome the drug's depression of network representation of self-awareness. We know that struggle goes on, and it even gets expressed in physical thrashing about. That is why orderlies strap surgical patients to a gurney and "hide" them in a recovery room while the brain is fighting to wake up. There is also a lot of poking and prodding stimulation during recovery from vital-sign checking, clean up, and bandaging.

Also, dreams and hallucinations may well be occurring during recovery from anesthesia that can't be remembered because the drug prevents consolidation of the memories. The other thing about anesthesia is that it is not the normal unconsciousness of sleep. There is no equivalent to SWS. The drug suppresses neural activity, particularly in the brainstem reticular formation and cerebral cortex. Also, blood flow to the cortex is reduced. These constraints on consciousness go away as soon as drug levels fall below their effective threshold.

Narcolepsy is the other problem for the HD theory. Narcolepsy is a disease, one in which the brain has sudden attacks of what appears to be dream sleep. How can this happen without being preceded by the deep slow-wave stage of sleep? First of all, in narcolepsy the brain goes from one state of consciousness to another consciousness state. The point is that both normal consciousness and dreaming are not so different that it would be difficult for the brain to transition from one state to the other.

I haven't yet addressed the issue of why the brain has more activated sleep when it has been deprived of it. In sleep labs, experiments have shown that waking a person up every time an activated episode begins will cause such sleep episodes to be much more frequent during subsequent nights, if they are left undisturbed. A similar effect can be produced by cessation of sleeping pills that suppress activated sleep. My explanation is that more than the usual amount of activated sleep is needed after deprivation because the brain is short on practice at waking itself up. This relates to the idea that rebooting wakefulness from scratch is a complicated task that, like a golf swing, requires perpetual practice.

We do know that when people are prevented from having their dream episodes, their conscious state becomes highly disturbed, even to the point of hallucinations. This observation reinforces the notion that dream sleep is central to the process of regenerating a cognitively competent conscious state as it emerges from the disruption of sleep.

Can the study of brain function in humans with incompetent consciousness, such as schizophrenics, help in understanding consciousness? This question comes to mind because a leading theory of schizophrenia is that dreams intrude into conscious awareness. What we do know is that drug-free schizophrenics tend to have insomnia, but they otherwise have sleep patterns, including activated sleep, that are similar to the sleep patterns of normal people.[44]

To conclude, let us recognize that there are incidental benefits of activated sleep besides the need for recovery from SWS. Just getting ready for the next day's conscious activity has its value. There is also benefit in consolidating certain memories that perhaps cannot be as effectively consolidated during other stages of sleep. Then there is the likelihood that dreams can have reward properties. I truly believe that my dog sleeps so much because she is bored and looks forward to the adventures, such as chasing deer and catching critters, that she has learned she can experience in her dreams.

Dreams are nature's way of enhancing the life experience. This also relates to a basic propensity of brain: stimulus seeking. Advanced animals have an evolved brain that feeds on stimulus. Activated sleep helps to address that need.

And what about my dog's dreaming? I hope she catches that damned squirrel.

CHAPTER 4

DOES CONSCIOUSNESS DO ANYTHING?

WHY ARE WE CONSCIOUS?

Astonishingly, a growing number of scholars assert that consciousness cannot do anything. If it doesn't do anything, consciousness can't have a purpose. It is regarded as a mere observer of life as it is played out on a Cartesian stage. All intentions, plans, and actions are generated unconsciously, with a portion of those processes made available after the fact to consciousness.

Such a notion flies in the face of common sense. But, of course, scholars delight in discrediting common sense. We can bolster common sense explanations by stating the function of consciousness in neural terms. Brain activity causes other brain activity that results in intentions, decisions, choices, plans, and assorted behaviors. Consciousness arises from and is part of brain activity. Therefore, the brain activity of consciousness should be able to cause and modify other brain activity.

There is another argument. Consciousness is an evolved brain capability found only in higher animals, most notably in humans. There are obvious advantages for the natural selection of a consciousness that can do things. If consciousness does nothing, why did it evolve?

Brain and body properties typically evolve because they satisfy some evolutionary advantage. Yet in his book *Mind and Consciousness* Thomas Nigel uses the premise that consciousness could not have evolved because the forces of natural selection have nothing to select for, since consciousness doesn't do anything. Without any value of consciousness, why would it be favored in natural selection? Thus, he has no explanation for the existence of consciousness.

Of course we could use the appendix argument. Not everything in human biology has to have a purpose. Some genetic coding is just carried along in the stream of evolution. But ask yourself this: Would we be just as functional as we are now if we were unconscious zombies or robots? Which is more adaptive: to evolve capacity for a false belief in our powers of introspection and consciously directed agency or to actually have such powers?

We should add that a conscious mind that does nothing useful is grossly wasteful of neural resources. The brain has to commit a huge proportion of its circuitry just to sustain consciousness, resources that otherwise would be available for unconscious processing. I see consciousness therefore as a brain-function add-in, one that is worth the diversion of computational resources because conscious thinking is a more effective way of performing certain mental functions.

A likely reason for scholars to think that consciousness has no purpose is the long-standing disdain for dualism, the notion that there is some kind of out-of-body spirit or force that directs behavioral choices and actions. Yet dualism is not necessarily relevant here.

The obvious effect of consciousness is that sensations and memories are made explicit. So the question becomes: What is the value of having such information processed explicitly in the brain? Surely, consciousness might, through its explicit restructuring of its worldly representations, provide another way for the unconscious mind to be taught and programmed, enabling better performance in the future. Also, the conscious rehearsal of learned information surely contributes to the consolidation of the unconscious mind's memory, especially when explicit associations and mnemonics are included in the effort to memorize. Eventually, with sufficient conscious rehearsal, the learned skills become incorporated into unconscious mind as habits, learning and memory schema, and motor skills.

Another obvious matter to consider is this: Why is there no agency when we are asleep or otherwise unconscious? If conscious mind can't do anything, why does agency only occur when we are conscious? The contrived answer might be that unconscious mind can't direct bodily action when it is asleep. If we could wake up without consciousness, functioning like zombies, then it would prove that consciousness is not needed to cause behavior. Maybe sleepwalking is pertinent here: sleepwalkers are awake enough to move around and do things, but not complex or insightful things. I should add that all humans act like zombies on occasion, even when awake, but for most people that state does not dominate wakefulness.

Common experience teaches that agency does not occur in unconscious states, such as sleep, coma, or anesthesia. If agency only occurs during consciousness, how can one assert with any assurance that consciousness is not relevant to agency and therefore has little purpose in its existence?

The conscious mind can create new understandings from experience. Thus, we can affirm the interaction of mind, body, and environment. We also have a basis for affirming the important role of learning in the development of the brain and the mind. Not all learning has to be conscious, of course. Indeed, implicit, unconscious learning is a well-established phenomenon in both animals and humans. We have no good way to know how much of our learning and brain processing is unconscious, though the prominent cognitive neuroscientist Michael Gazzaniga makes the claim that 98 percent of what our brains do is unconscious.[1] Where does he get such a number? This is reminiscent of the widely accepted myth that we only use 10 percent of our brain capacity. Nobody can show where or how such numbers are derived.

Religious arguments do not help much in trying to understand the value of consciousness. But let us not assume that the material basis of mind is necessarily incompatible with religion. Science and religion seem separated by a territorial line, but in reality this line may be a broad, fuzzy landscape that can be crossed from either side. E. O. Wilson, in his book on consilience,[2] said that the misunderstandings arise not from fundamental differences but from ignorance of the fuzzy boundary.

Religious people, in particular, are attracted to Wilson's view because it allows for the idea of a nonmaterialist soul, which they think is a mandatory part of their belief system. Descartes was among the first to assert that the soul cannot have material properties. There is no evidence for such a claim. Moreover, a close reading of comments attributed to Christ and to St. Paul, for example, revealed a claim that in the afterlife the soul does have its own body, albeit one that differs from what we know about bodies (see if this applies in chapter 5).

THE VALUE OF CONSCIOUSNESS

When you wake up in the morning and fix your breakfast, you not only know you are fixing breakfast and fixing it for yourself, but you know you know that. You know who you are and you know the role that you are playing in this breakfast-fixing scenario. In the process, you make conscious decisions about what to fix and how to prepare it—all in the context of self. Your conscious mind appears to plan a desired sequence of events: perk the coffee, take pills with orange juice, get out the right dishes and utensils, and so on. Your conscious mind can veto decisions. For example, you may choose to skip the eggs because you have learned that too much cholesterol is bad for you.

Social interactions are enriched when we are conscious—that's certainly valuable. When conscious, we have an explicit way to recognize the existence of conscious mind in fellow humans. Consciousness likely provides a magnifier effect for the mirror-neuron function described in chapter 3. Thus, we can anticipate in explicit ways what others are probably thinking, predict what they will do, and facilitate any similar actions we might take. Thus, humans can interact with each other in more appropriate, and more complete and productive ways than they could if their unconscious minds had to figure out what other unconscious minds were thinking.

Introspective awareness also occurs in consciousness. Introspection magnifies the relevance of sensations, thoughts, and feelings. Introspection also allows the brain to live vicariously in the world of imagination, where thought need not be directly tied to sensation.

There are those that assert that what we consciously note as our "self" is only a model generated by an unconscious part of the brain and then made available after the fact to consciousness. Sense of self, even in unconscious form, is a real sense, no less so than traditional senses. It is clear that the self/non-self distinction begins forming with the fetus's topographical mapping of the body. But since in mature brains the model seems processed explicitly in consciousness, the model must be reconstructed in consciousness. This goes to the argument of whether or not consciousness does anything. But what if consciousness is itself an amalgamation of neural events? Neural events cause other events. Thus consciousness can not only perceive neural events, but also produce them.

There's more. I think that neocortical executive control is enhanced by the explicit awareness, introspection, and analysis afforded by consciousness. Yet this rather obvious conclusion has been recently challenged, on the basis that unconscious processes can trigger or even participate in executive control.[3] But where is the evidence that the two possible sources of executive control are mutually exclusive?

Such crucial processes as information organization, reflection, analysis, planning, and executive control occur during consciousness. Do these processes not impact what is going on unconsciously? It seems natural to suggest that the main function of consciousness is to refine, amplify, and perfect unconscious processes. If consciousness is only a display mechanism, then of course it could not do any of the things we usually attribute to it.

Consciousness is a quality-control mind, one that can improve the quality and effectiveness of our thinking and behavior. We use conscious mind, particularly its power of language, to program our brain via the analyses, decisions, and choices we make. Conscious mind accelerates our learning and development of competence. Conscious mind fine-tunes our belief and value systems.

Awareness of one's thoughts gives the brain a better chance to know what its own processes are doing, at least in terms of planning and awareness of the consequences of those processes. Such awareness provides for refereeing, editing functions, and ongoing guidance of ongoing actions. Efficiency as well as effectiveness of thinking is enhanced when such explicit awareness can be used to supplement what is going on unconsciously. Thus, we can see why humans are so well adapted for life in a complex and difficult world.

AGENCY

By *agency* I mean the intentional effort to form and recall memories, evaluate values, conduct deliberations that lead to decisions, and plan in an organized way that produces overt action. While many such processes no doubt occur unconsciously, they are of most interest when they occur during consciousness. Each of these elements of agency is distinctly different, and each no doubt has its own unique circuit impulse pattern (CIP) representations.

Two key principles of agency seem evident:

1. Agency arises from a nervous system complex enough to have a sense of self. In lower animals, this sense is necessarily unconscious.
2. Agency processes are nonlinear. Indeed, most everything a nervous system does is nonlinear.

So why does nonlinearity matter? In linear systems, any response is directly proportional to changes in its causes. If you graphed such changes, you would get a straight line. Such a process is highly predictable. Nonlinear processes produce cause and effect relationships that are disproportionate. Over the long term, the effects may be best described by the mathematics of chaos theory.

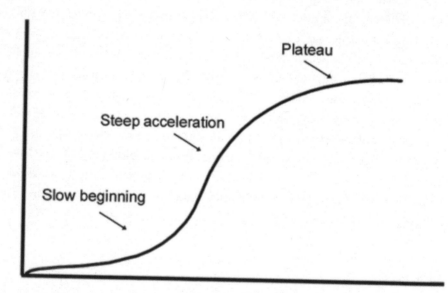

Figure 4.1. Cause-effect relations in living systems. Typical cause (horizontal axis) and effect (vertical axis) relationship curve for living systems. Such relationships are typical at all levels, including cellular biology, bodily physiology, and brain functions. For any given phenomenon, the specific level of effect is much harder to predict than in simple linear systems.

Over the short term, the brain's nonlinear processes typically produce a slanted S-shaped curve, reflecting a small response to low levels of the cause (such as stimulus or dose of a drug), a rapid acceleration of responses at moderate levels of cause, and a plateau where increases in the cause produces little effect (see figure 4.1). Very large changes in cause level may cause a catastrophic sudden drop in response, as in death.

I have not seen attempts to apply this principle to understanding agency. But what it suggests to me is that for any element of agency (intention, memory recall, and so forth), a low level of effort may produce little or no effect, a high level of effort will produce diminishing returns with increasing effort, and accelerating consequences occur over a mid-range of effort.

Ultimate behavior results from the accumulation of causal processes of each element, though each element of agency may have its own cause-effect relations. For example, formulation of intentions drives the recall of relevant memories. These in turn generate relevant emotions and assessment of relevance, value, and anticipated consequences. Decisions are made on the basis of foregoing processes, and planning develops as appropriate to the decision. Then, and only then, do you get a final result. No wonder that human actions are so complicated, often unpredictable, and often undesirable.

Fortunately, the brain has a system of executive control to help regulate these elements of agency, which left to their own devices could run amok and produce catastrophic effects. Brain-scan and electrical-recording studies have identified in general where in the brain this executive control resides: in the dorsolateral prefrontal, parietal, and cingulate cortices.[4] As you should expect, deficiencies in the function of this network underlie numerous neuropsychiatric conditions.[5] The ability to regulate emotions and direct rational actions is typically associated with success in life, and inability to do so often leads to dire consequences.

One seldom-discussed aspect of agency is initiative. Why are some people active while others are passive? Why are some people self-starters? Some of the difference is no doubt genetic. But surely what we experience, especially while growing up, has a lot to do with it. And then, if it exists, there is free will (see below).

If a child has a lot of negative experiences growing up, that may surely teach the child to be hesitant, to be less adventurous, and to be risk averse—

all of which would stifle initiative. Much of this conditioned tendency would be learned unconsciously. But conscious introspection can be used to counter such conditioning. Wall flowers do sometimes turn out to become the life of the party.

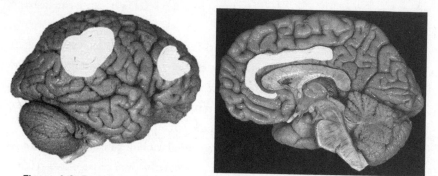

Figure 4.2. Brain executive control. Approximate regions of the brain that constitute an executive control system (shaded in white). *Left:* the lateral side of one cerebral hemisphere showing the dorsolateral prefrontal cortex and the parietal cortex. *Right:* anterior cingulate cortex.

A person who wants to do something about his or her lack of initiative should ask, "If I can do it, should do it, and want to do it, what am I waiting for?" Many of our perceived failures in life are not created by what we do but by what we fail to do. Maybe we just procrastinate and don't get around to action. Or maybe it's just a thought, something that we think would be nice to do, but we just aren't serious about it.

What keeps us from action? Can, should, and want ought to be pretty compelling. One excuse is that we just can't seem to find the time. That won't wash. Whatever we do in life, we have found or made time for. Final choices are matters of priority, and sometimes we don't prioritize well.

Fear is an obvious cause of passivity. There are many kinds of fear that cause inaction. There is

- fear of failure,
- fear of being different or out of step,
- fear of rejection,
- and even fear of success.

Fear of failure arises from self-doubt. We may think we don't know enough, don't have enough time or energy, or lack ability, resources, and help. The cure for such fear is to learn what is needed, make the time, pump ourselves up emotionally so we will have the energy, hone our relevant skill set, and hustle for resources and help. These things can be demanding. It is no wonder there are so many things we can, should, and want to do but don't.

All our life, beginning with school, we are conditioned to consider failure to be a bad thing. But failure is often a good, even necessary, thing. The ratio between failures and successes for any given person is rather stable. Thus, if you want more successes, you need to allow for more failures. Two-time Nobel Prize–winner Linus Pauling once said, "The way to get good ideas is to have lots of ideas."[6] Even the corporate world recognizes this principle, and the most innovative companies practice it. Jeff Dyer, in his book *The Innovator's DNA*, says the key to business success is to "fail often, fail fast, fail cheap."[7] It's okay to fail, as long as you do it fast and cheap and learn from it. Our mantra should be: "Keep tweaking until it works." This is exactly how Edison invented the light bulb. Most other inventors and creative people have operated with the same mantra.

Fear of being different stifles initiative. It can cause people to join groups, causes, and lifestyles that are not good for them or even harmful. The corollary is that bad social commitments make it harder to experience better alternatives. Not everyone can be a leader, who by definition is different from the crowd. But all of us are better off when we strive to become our own person.

Fear of being different often arises from personal insecurity and lack of confidence. These are crippling emotions, and one's life can never be fully actualized until they are overcome. This comes to the matter of self-esteem. One thing many people don't realize is that self-esteem has two quite distinct components: self-worth and self-confidence. Self-worth is given (by being valued and loved by others, or by a belief in God's love). Self-confidence cannot be given—it has to be earned. This is a major reason why peer pressure drives so much of teenage behavior. At this age, young people haven't had much chance to consolidate personal identity and develop confidence. People who lack the confidence to "put themselves on the line" deny themselves opportunities to enjoy the fruits of success. Their life becomes a vicious cycle that begins with lack of confidence, lack of agency, lack of success, and increased justification not to be confident.

If we are different, the in crowd may reject us. Rejection is certainly depressing. Nobody in his or her right mind wants to be depressed. But no life can be fulfilling when it is lived to satisfy the opinions others may have of us. We need to be true to ourselves, to trust in our values and standards. If such trust is not justified by who we are, we can certainly fix that. This dictum lies at the heart of Socrates' great admonition: "The unexamined life is not worth living."

Fear of success is often learned by watching how others have failed to adjust to success. Witness the entertainment celebrities who end up committing suicide or devolve into unlikable people. Most of us probably know personally some people who have become conceited, aloof, condescending, arrogant, or otherwise unlikable as a result of their success. We don't want that to happen to us. When we surrender to our fear of success, we affirm our lack of trust in ourselves. Do we really need to reinforce such lack of self-trust?

This brings us back to the question of why the human brain evolved the capacity for consciousness. Acts of agency are most effectively accomplished in the explicit light of consciousness and introspection. At issue is whether agency is produced by conscious processes or just recognized consciously. The scientists who contend that consciousness is worthless and does nothing (see for example Haggard[8] and Wegner[9]) say that neural events cause agency and consciousness, but contend that consciousness cannot cause either neural events or agency.[10] They believe that consciousness is an epiphenomenon of brain function and, as such, can only make explicit some otherwise unconscious thought. In other words, consciousness can observe but not do. I have found no actual evidence for such a view, other than speculative extrapolation from simple reflex-like movements purported to disprove the idea of free will (see below). But even if one accepts the position that free will is illusory, how does that inevitably support the position that there is no conscious will and action at all?

Later in this chapter, after considering free will, I will summarize what I think conscious mind does.

THE BRAIN'S AVATAR

To explain what causes or constitutes consciousness we have to clarify how we think about consciousness. For example, Michael Gazzaniga, famous for studies of split-brain patients, postulates that consciousness has a locus in the brain, probably in and around the left hemisphere's language centers.[11] Thus, his idea is that consciousness is language based; that is, he considers consciousness to be a kind of verbal interpreter. Yet we know that people who lose their language capabilities through stroke are still conscious. Gazzaniga's own experiments in split-brain patients show that the right hemisphere is consciously aware of input it receives even though it has no language-generating capability to describe what it perceives.

I have a different take that emphasizes basic neurobiology and minimizes the usual speculations of psychology and philosophy.[12] The interpreter idea is only one of several possible explanations, and it does not address mechanisms. Just how does consciousness exist in the brain? In chapter 3, I explained how neural circuit impulse patterns (CIPs) underlie all basic brain functions, including the state of consciousness.

Such a state of awareness (knowing that you know) exists as a set of CIPs that engages CIPs in areas outside the primary sensory pathway and collectively enables agency. For example, the CIPs in the sensory pathway during consciousness may become coupled with CIP processing that engages circuitry for emotional response to pain stimulus, to memory stores of similar past painful events, to stress-hormone pathways, and to motor-control pathways that can direct withdrawal movements.

This CIP state of consciousness is contained within the metacircuit of the whole brain and is therefore integrated with what goes on in unconscious processing. The CIPs of conscious self are the equivalent of a brain-created avatar that acts in the world on behalf of the brain and body. I suggest that it is these conscious CIPs that make us *human* beings.

People who play computer games know about avatars. These are computer-generated proxies for the gamer. The avatar is the game player's agent, doing things in the game in response to commands from the player. A good example is the increasingly popular Web environment known as Second Life, in which players create their own avatars and live vicariously through the avatar in the virtual world. The analogy of computer avatars breaks down when it comes to human consciousness avatars. Computer avatars cannot generate their own agency.

Brain avatars respond to impulses and instructions from unconscious mind as well as ongoing environment events. Unlike computer avatars, however, the brain's avatar may act on its own initiative to serve the best interests of the brain and the body. Perhaps the avatar has an interest of its own, a mind of its own, so to speak. Moreover, the biological avatar gets to decide, or at least influence, what the brain thinks is in the best interest, and it supervises the actions to accomplish it.

This avatar notion is reminiscent of an old third-century idea that there is a little person inside the brain that is a scale model of the body. In modern terms, however, this *homunculus*, as it came to be called, is used as a way to think about how the body is mapped in the sensory and motor cortex. But the homunculus is more than that. It is a CIP representation of the body and what goes on inside and outside the body, all referenced to the sense of self, which itself is a CIP representation of the "little person." I contend that the brain creates a conscious homunculus in the form of an avatar, which it deploys to act on behalf of the embodied brain in ways that would not otherwise be possible if the unconscious mind were solely responsible for human behavior. The rough outline of each person's homunculus was constructed in the womb, as was the neocortical circuitry for accomplishing consciousness. However, humans construct the content of consciousness continuously throughout adult life.

Human avatars should have the capacity to sense, intend, evaluate, decide, plan, and direct action. What would be sensed by the avatar? First and foremost, because the avatar thinks consciously of itself, it constitutes itself as an avatar knowing that it acts to benefit brain and body. The point is that thinking of oneself as an avatar and being an avatar are equivalent. Recall the famous saying of philosopher René Descartes:

I think, therefore I am.

This being is the conscious sense of "I." It senses much of what the brain is thinking, such as beliefs, wishes, decisions, plans, and the like. Moreover, consciousness can sense how it is teaching the unconscious brain in terms of specific cognitive capabilities, motor skills, ideas, attitudes, or emotions.

The human avatar thinks of itself in some of the biological ways that subconscious and nonconscious minds think of themselves (as existing as part of the embodied brain). But the avatar thinks of itself consciously.

Any time we are awake, the avatar being is active, deployed on-line so to speak. By way of computer analogy, when the avatar is "on-line" during wakefulness, it is operating in RAM and able to exert its functions. When the avatar is shut down, as in going to sleep for example, the avatar goes off-line and saves its CIP files on hard disk. In biological systems, the hard disk is stored in the neuron terminals and synapses of the preferential segments of the global neural network that holds the memory of self and the capacity for rebooting the self. The self may have undergone some subtle changes with the day's experiences, and updating the modified self in long-term memory is one of the functions of sleep, as was explained in the previous chapter. That chapter also explained the rebooting of the avatar after sleep.

The really hard question is, how can such an avatar exist as a conscious being? What is it about the CIPs of the avatar that empowers it to evaluate input in the context of a conscious self-awareness? Nobody knows, but I will speculate that the consciousness exists because the avatar was created as a "second self-aware self," which is a necessary condition for consciousness.

The first part of my explanation lies in the fact that the conscious state makes us explicitly aware of ourselves as the "I" who sees, hears, and so on. So, the important question is, how does the brain make us explicitly aware of ourselves and our engagement with the world? Answer that and all else follows.

When engaging in introspection, humans have a distinct awareness of the nature of thought but no inkling of the CIPs that represent those thoughts. That is, we are aware of the representations but not their codes.

Why aren't we aware of the combinatorial CIP codes? The brain thinks via the codes, but our consciousness only makes us aware of the codes' representations. That is as it should be, because it is the representations that we engage with in the real world in terms of what we see, hear, or otherwise sense and remember.

The avatar's agency will have to have something to do with how the CIPs in the brain's topographical maps are accessed. The mapped CIPs must be bound dynamically in a way that makes their representations accessible in a conscious state. Or is it more precise to say that once a conscious state exists, CIP representations can become explicit?

But how does a CIP-based avatar receive its conscious self-awareness? Consider this example: If the CIPs in two interconnected cortical columns become time locked, they are sharing CIP information in an especially intense and perhaps magnified way. Maybe this is all it takes to lift the representations of those CIPs into explicit realization, assuming that many cortical columns across the global workspace are time linked. Or does a pre-existing state of consciousness allow explicit realization of the CIP representations if and only if the specific CIPs of a given representation are in a specified time-locked phase relationship? Recall the earlier discussion of how a state of consciousness is created by brainstem reticular formation activation of the neocortex. Better stated, it is the reticular-formation drive that achieves the temporal linkage of cortical circuits needed for explicit realization of specific CIP representations that are brought into consciousness.

In the unconscious state, CIPs can maintain their sensory and remembered representations without having a mechanism for explicit awareness of them. In the conscious state, the CIPs across the brain's global workspace develop a coherence of impulse activity that enables explicit awareness. The enigma lies in explaining the nature of the CIP coherences that enable explicit awareness.

In consciousness states, the amplification of CIP information content may produce two data streams, one the default information of a single-stream CIP representation that resides in unconscious mind, and a second data-stream duplicate that is processed in a time-linked but slightly delayed fashion with the first. The corresponding mental representations likewise exist as two coherent time-delayed streams, and it is the confluence of the duplicate data streams that constitutes consciousness. This might help explain why conscious realization of a thought or experience is slightly delayed, which is a fact that is part of my criticism below of experiments purported to prove that free will is an illusion.

Moreover, we should consider the possibility that consciousness arises not just because of the amount of engaged circuitry but perhaps also because of a unique way in which the circuits are engaged. Much current research shows that the degree of synchrony and time locking of CIPs in various regions of the cortex is associated with conscious processes. Moreover, a few experiments report shifting phase relationships of activity at different frequencies within the same brain area. The oscillating field potentials that are associated with impulse activity in a given area can occur in two or more frequency bands, and their phase relationship should surely be consequential. Depending on the nature of stimulus and mental state, these oscillations may jitter with respect to each other or become time locked. The functional consequence has to be substantial, and I suggest that this is a fundamental aspect of consciousness.

Let us rethink in an avatar context what consciousness is. Awareness means that an awareness "target" CIP representation has been created for a sensation or a thought or a memory, similar to the process that occurs even in unconscious mind. During conscious awareness, the avatar CIP must "read" the message in the target CIP and thus in that sense be aware of the target. The "reading" may occur inside the avatar's own CIPs. Since the avatar knows who it is from the CIP representation of its sense of self, it simultaneously knows that it knows about the target CIP. This representation can be compared and evaluated with representation of other targets as they are experienced or recalled from memory, all of which can be processed in the same avatar CIP environment. Such a process may not be much different from what must occur when unconscious mind processes new information and integrates that with memory stores.

How could such an avatar do things? Most of us assume that our avatars are not only self-aware but also make decisions to act on behalf of our brain. How would that be accomplished? Why not consider that "little person" viewing the stage of a Cartesian Theatre as more than a metaphor? The "little person" could be viewed as a neural representation of self that can not only view some of what the unconscious brain is doing but also act on it. Because the avatar is actually a set of CIPs interacting with other CIPs, it can modify and be modified by what is happening in the other CIP sets.

When the brain constructs a CIP representation of a sensation like sound or sight, as far as the brain is concerned, the representation *is* the sensation. It is the representation that the brain is aware of, not the outer world as it really exists. Remember the explanation I gave in chapter 2 for how the brain abstracts and deconstructs images and represents them as CIPs. When consciousness is present, its CIP set is the *representation* that the conscious mind is aware of.

Another way to say this is that the avatar is aware of the CIP representation of the sensation of consciousness as well as that of the target experience. So you have one set of CIP representations, of the unconscious sensations and processes, sharing CIP information of another set, the avatar.

If consciousness is a code of CIPs, how can such a code create personality, beliefs, feelings, decisions, and so forth? This question is wrongly posed. The code may help create these things, but it is also the representation of such things.

The code, because it arises from materialistic CIP processes, is an essential part of the machinery of mind. The code can therefore influence the very circuits from which it is being generated. This is key. Read it again, if necessary. In that way, a code in real time can change the nature of the code at some future time. In short, conscious mind can change its mind. If the code is replayed and rehearsed enough, the mind change becomes a permanent part of long-term memory.

Let us apply this avatar idea to explain how conscious mind operates. Brain creates the avatar, and no doubt much of the control of the avatar is exerted by the brain's unconscious mind. But the brain's avatar may be granted a degree of "mind of its own." That is, the brain avatar may have some freedom to analyze, intend, and make decisions on its own. In any case, the whole point is that the avatar serves the interests of the brain that creates it. The beauty of this arrangement is that the avatar, being conscious, can do things the unconscious mind cannot do by itself.

My idea is that the brain creates a unique set of CIPs that represent a "sixth sense," a conscious sense of self, which because it arises from neural circuitry can readily interact with unconscious mind and with the external world—and *do* things!

This CIP avatar is an explicit surveillance agent that monitors the CIP information that comes from unconscious mind and from the external environment. Sounds like the Cartesian Theatre's "little man" discussed above, doesn't it? One difference is that the avatar does more than just watch what goes on. It is an active agent of the brain, acting on behalf of the interests of the body and the brain.

How could the avatar's CIP representations monitor what is going on in the brain, as in the case of sensory stimuli, for example? One possibility could be that the conscious mind does not "see" original stimuli, but mainly "looks in on" the CIP representation being held in the unconscious mind. Certain sensations, such as itch or pain, can only be perceived in the conscious mind.

Representations of such stimuli are contained in CIPs, but when accessed by the CIPs of conscious mind, these representations are detected with the special conscious identity of either itch or pain. Conscious mind (that is, the avatar) probably contains CIP representations of another sort. Namely, the brain creates a conscious mind that is a representation of self-identity, as distinct from representations of the external world.

I would put it this way: consciousness is not something "out there," but something (i.e., CIPs) "in here." In other words, the so-called ghost in the machine is not a ghost at all. "I" really have a physical (CIP) existence.

Now the really hard question is, how does the avatar operate consciously? No one has figured that out. Conscious mind is an entity, in the sense that it is a neural representation of self, one that is grounded in the currency of all thought, CIPs. The hardest part of this hard question is, why should this sense of self be conscious? How does it accomplish a level of knowing beyond the mere detection that operates in unconscious mind? Indeed, even the unconscious mind must reference its operations to the bodily sense of self. We must ask not only how the avatar is aware of its existence as a unique entity, but also how it knows that it knows what the other five senses detect?

Advanced nervous systems still use the same currency of thought that primitive systems use. What, then, could be different about the neural activity of consciousness? First of all, we have to identify the neural correlates of consciousness, as many neuroscientists are now trying to do. Further, we must identify which of these correlates are essential causes of consciousness, rather than mere coincidental correlates. I suggest that science has not completed this task, inasmuch as combinatorial codes have not been sought and temporal coherences have received only modest inquiry. Both mechanisms most likely increase the informational "carrying capacity" needed to achieve the more robust processing capacity of consciousness. Carrying capacity is a central requirement of consciousness. Conscious thought must operate on small chunks of information at any given instant (see my comments elsewhere on working memory).

Clearly, this must have something to do with the fact that the avatar, by definition, processes information in the context of its own self-identity. It is the avatar that is self-aware. The nervous system's fundamental design principle is to accomplish awareness—to detect things in the environment and then generate appropriate responses. In higher animals, that capability extends to detecting more and more abstract things, ultimately the most abstract thing of all, the sense of self. Such a CIP-based system is not only able to detect and code events in the "outside" world, but it can do the same for its inner sense of self. Thus, the conscious mind, being simultaneously aware of the outside world and its inner world automatically has the capacity to know that it knows. This sense also has an autonomy not found with the traditional five senses. It is an entity that has a life of its own.

The circuits of the avatar overlap those of the senses and their unconscious processing circuits. Thus, the information content in the circuits mediating traditional senses is accessible by the avatar circuits. And when that information is accessed, it is processed in a self-aware system, in turn making explicit some significant portion of the traditional senses and their associated thought.

The CIP theory regards consciousness as a global-system property, involving dynamic and reciprocal interactions between certain mostly cholinergic neurons in the brainstem and basal forebrain, the reticular zones of the thalamus, and widely distributed but interconnected cortical columns of species with advanced neocortical architecture. Moreover, important properties of consciousness derive from the synchrony of neural activity of various frequencies in multiple cortical zones.

The CIP view holds more specifically that the internal representation occurs in the form of a select global set of neural circuit impulse patterns that collectively constitute conscious personhood in a form equivalent to a virtual avatar that acts on behalf of the brain that generated it. The avatar acquires modification through life experience so that the specific details of one's conscious personhood change over time. When "off-line," as during sleep or anesthesia, that avatar is stored as memory in facilitated synapses of the circuits that can regenerate it upon awakening.

The conscious self is an explicit frame of reference by which all senses can be evaluated in the context of self-awareness. The smell of sizzling steak is only *detected* by my olfactory pathways; it is *perceived* by my "I." Likewise, the sight of a beautiful woman, or a touch of kindness, or the sound of great music, or the taste of fine wine; these are all detected by my primary sensory pathways but perceived by my "I," my conscious sense of itself. The avatar is what makes us human.

What about abstract thinking? This requires consciousness, indeed a very high form of consciousness. If consciousness is an avatar, itself an abstract representation, then you have a mechanism of consciousness in which CIPs that represent the avatar are able to create other CIPs that are abstract representations of thought.

Maybe I am a figment of my brain's imagination . . . but wait: I have to be more than a *virtual* me. I am created and represented in the form of real brain circuitry, in the wetware of nerves and aqueous solutions. When "I" am on-line, I exist as patterns of nerve impulses propagating throughout that circuitry. When I am asleep I exist as preferred junctions among neurons that store me on the biological equivalent of a "hard drive" that has the capacity to put me back on-line. Remember, all these processes operate throughout our brain's odyssey from womb to tomb.

Avatar Supporting Evidence

There are lines of evidence that support the CIP theory in addition to the rationale just developed. Evidence falls into two categories of predictions: (1) the CIPs, or some manifestation thereof, such as EEG frequencies, should change as the state of consciousness changes, and (2) changing the CIPs or their manifestation should change the state of consciousness.

In the first category, the whole history of EEG studies, both in laboratories and in hospital settings, attests to the fact that there is generally a clear correlation between the EEG and the state of consciousness. Note that the signal source for the EEG is the part of the brain closest to scalp electrodes, which is the neocortex. There are apparent exceptions, but I have argued elsewhere that these EEG-behavioral dissociations, as they are called, are usually misinterpretations of the state of consciousness.[13]

The general observations can be summarized as follows:

- In the highest state of consciousness and alert wakefulness, the EEG is dominated by low-voltage, fast-activity (beta and gamma) waves, typically including oscillations in the frequency band of forty and more waves per second.

- In relaxed, meditative states of consciousness, the EEG is dominated by slower activity, often including so-called alpha waves of 8–12 per second.
- In emotionally agitated states, the EEG often contains a great deal of 4–7 per second theta activity. In drowsy and sleep states, the EEG is dominated by large, irregular slow waves of 1–4 per second. The large, slow electrical waves that permeate the neocortex during unconsciousness presumably reflect synchronous activity of neurons that are causing the inhibition that prevents emergence of arousal and consciousness.
- In a coma state, the trend for slowing of activity continues, but the signal magnitude may be greatly suppressed, even to the point of the signal being undetectable from the scalp.
- In death, there is no EEG signal anywhere in the brain. There is no EEG because neurons have stopped firing their CIPs.

Because the EEG is a manifestation of overall CIP activity, changes in EEG correlates of consciousness support the notion that it is changes in CIP that create the changes in states of consciousness and in the corresponding EEG patterns. Even so, these are just correlations, and correlation is not the same as causation.

More convincing evidence comes when one can show that changing the CIPs, either through disease or through some external manipulation, changes the state of consciousness. For instance, massive cerebral strokes may wipe out neural responsiveness to stimuli from large segments of the body, causing the patient to no longer have any conscious awareness of stimuli from such regions. Injection of a sufficient dose of anesthetics produces immediate change in neural activity, and unconsciousness ultimately follows. Similar effects can be produced unilaterally by injecting anesthetic into only one carotid artery. Naturally occurring epilepsy causes massive, rapid bursts of neural activity that wipe out consciousness. Even during the "auras" that often precede an epileptic attack, there are localized signs of epileptic discharge and the patient may be consciously aware that a full-blown attack may soon ensue.[14]

Another line of evidence comes from the modern experimental technique of transcranial magnetic stimulation. Imposing large magnetic fields across spans of scalp is apparently harmless, and doing so produces reversible changes in brain electrical activities that in turn are associated with selective changes in conscious awareness. A wide range of changes in consciousness functions can be produced depending on the extent of tissue exposed to the magnetic field stimulus.[15]

Summarizing, let us first remember that wakefulness has to be triggered, typically by activation (actually disinhibition) of the neocortex from the brain-stem reticular formation and its reticular thalamic extension. EEG signs of wakefulness do not necessarily equate with consciousness. The EEGs of various lower animal species all show an "activated" EEG. I surveyed these reports years ago when I was asked to write a section of *Biology Data Book*.[16] Such data, and the conclusions based thereon, were obtained before the age of digital EEG and frequency analysis. The pen and ink tracings of older EEG machines did not effectively display frequencies above about 30 per second. What have not been studied in lower animals are the frequency bands and spatial coherences of such EEGs. It is entirely possible that lower animals only have beta activity (less than about 30 per second), while conscious humans have more predominance of gamma activity (40 and more per second). Moreover, the degree and topography of coherences have never been subjected to examination across species. One index of degree of consciousness could well be the ratio of gamma activity to beta activity. Another index could be frequency-band-specific differences in the level and topographic distribution of frequency coherence. These animal studies really need to be performed again with modern electronics.

If it turns out that gamma activity and its coherences (both topographic and with other frequencies) are central to consciousness, we then must find an answer to the question of how the underlying high-frequency clustering of impulses that account for gamma activity could create a consciousness avatar. At this point, the answer is not clear. But, as mentioned in the oscillation section of chapter 2, the high frequency of gamma activity reflects more active neural circuits with larger capacity for carrying information and mediating throughput.

The CIPs of unconsciousness and consciousness might be basically the same, with the exception that consciousness might arise from some transcendent parameter, such as frequency phase locking. It seems likely that binding of CIPs and their disparate elements of thought is the major element of conscious thinking.

But it may be that it is not binding, as such, that creates consciousness, but rather the *kind* of binding. For instance, in my lab's study on ambiguous figures, cognitive binding was manifest in coherence in two or more frequency bands. These might even have had meaningful cross-frequency correlations, but that was not tested. Even so, it is not clear why or how consciousness would arise from multiple-frequency binding unless the coherence in different frequencies carries different information. One frequency may carry the information while another might carry the conscious awareness of the information. Another possibility is that coherence creates consciousness only if enough different areas of the brain share in the coherence.

Only a fraction of unconscious processing seems accessible to consciousness at any one time, suggesting that only a subset of CIPs could acquire the conditions necessary for consciousness. The corollary is that conscious registration has a limited "carrying capacity," and that does seem to be the case. Maybe this is because the CIPs of consciousness have to hold in awareness not only the unconscious CIP information but also the CIP information necessary for the sense of "I" and all that it entails.

Why Have Avatars?

This whole business of representing the physical world in such CIP representations as avatars must serve a purpose. There are certainly consequences of self-representation via conscious avatars, but you might think that human zombies could get along quite well without their avatars. Lower animals do quite well without conscious avatars. And humans are so smart, they would surely have survived as a species without being so conscious. A smart zombie should do quite well in terms of survival.

Indeed, you might suspect that our original primate ancestors of around two million years ago were just that: smart zombies. So now the question is, why didn't early hominids survive as such? The one apparent factor is that their brains were still not "large enough." Most likely, a bigger brain was needed to have enough circuitry to run an unconscious mind and simultaneously generate a self-conscious avatar. If those brains became successively more conscious over the eons of evolution, they could have been preferentially favored by natural selection.

But did we have to become so smart and so conscious in order to outcompete animal competitors who were surely less competent than today's monkeys and chimpanzees? Creatures like that weren't much competition for resources and survival. No, the real competitors were fellow subhumans, the less able zombies. There, the competition level was high, and a premium was placed on people who were smart and also conscious.

The fossil record clearly shows that there was a lot of killing going on among early humans. All of recorded history shows the same thing. Perhaps the human species is still being selected by the forces of war for more intelligence and more conscious thinking. Maybe we have reached a point where the selection pressure is now on the peacemakers who may inherit the earth left by the many millions who have died in human wars. Wars typically kill off the young before they have fulfilled their reproductive capacity. Evolution is all about who reproduces.

One major consequence of having conscious avatars is that they create religions, because the avatars believe that supernatural belief is in the interests of the self. The vast majority of humankind believes in some kind of creator god. All evidence indicates that this kind of belief existed even in the earliest humans. As cultures evolved, a variety of religions were fabricated, some quite elaborate, such as the pagan religions of the Egyptians, Babylonians, Greeks, Romans, Asian Indians, and Mayans. These religions were creations of the "I avatars" in the cultures from which they emerged.

One has to ask, what purpose is served when our avatars create religious beliefs? Religions, once formalized, do impose standards of belief and behavior. People conform to those standards or become ostracized. So religion has the effect of producing group mores, morals, and cohesion. Nicholas Wade wrote a whole book arguing this case.[17]

By definition, a conscious avatar has the capability for introspection about its existence, past, present, and future. Having the capacity to think about the future, avatars naturally do so. They know about death from seeing it all around them. It takes no great intelligence to know that they, too, will die someday. Why would we be surprised that they think about the possibility of an afterlife? At a minimum the afterlife idea can be comforting. Likewise in some schadenfreude sense, it is comforting to believe that there is an unpleasant hell for people the avatar does not like. So certainly religions may have been invented because avatars feel good about the idea, especially if they can construct the details on their own terms.

Believing that religion is a human creation is the basis for atheism, which basically is the belief that earthly life is all there is, that belief in anything else is a fabrication of the avatar imagination. That is not proof for atheism. Atheism, like religion, is a *belief* system.

How can a material mind have subjective beliefs? The avatar explanation of consciousness helps address the point raised by blogger Michael Egnor, who asserts that "matter, even brain matter, has third-person existence; it's a 'thing.' We have first-person existence; each of us is an 'I,' not just a thing. How can objective matter fully account for subjective experience?"[18] He tries to argue that "not a single first-person property of the mind—not intentionality, qualia, persistence of self-identity, restricted access, incorrigibility, nor free will—is a known property of matter."

But from an avatar perspective, these properties of mind do not have to have the properties of matter—the matter of CIPs *represents* those properties. If "I" am a CIP-based avatar acting on behalf of my brain, then my CIPs can include representations for the features of conscious mind that Egnor thinks are fundamental: intentionality, qualia, persistence of self-identity, restricted access, incorrigibility, and free will. Our CIPs, a materialist product of brain, acquire nonmaterialist properties when the representational CIPs are constructed as an avatar. The avatar CIPS are free to perform mental acts, such as free will, because they are not totally controlled by the constraints of unconscious mind. The avatar, though it may owe its birth to unconscious mind, is released to be its own active agent in the world.

One of the avatar's mental acts is to help construct one's sense of self.[19] Though this sense is laid down unconsciously in the womb, conscious life specifies and refines the SoS. A major purpose of the avatar is to help shape the SoS. It does this in a series of ways:

- Assessing self-worth.
- Choosing what the embodied brain will experience.
- Choosing what should be remembered.
- Interpreting experiences and memories.
- Deciding how to respond to experiences and memories.
- Responding to feedback from what we do.
- Forming a level of self-confidence.

Unleashing the Avatar

The avatar can only act when we are awake. Where does it go when it's taken off-line, as during sleep? Obviously, a stored form of the avatar exists that is automatically resurrected each day upon awakening.

How can the "I" avatar get unleashed from its slumber? There must be a threshold for the nonlinear processes that create the conditions for emergence of the avatar and its properties of wakefulness and consciousness. Though some people wake up in the morning more groggy than others, consciousness, for others, suddenly "comes on" like a light switch. Though we cannot yet specify these processes in detail, we know that the brainstem reticular formation interacts with neocortical circuitry to reinstate awakening, in the process unleashing the avatar.

How the Avatar Knows It Knows

How can the avatar be aware of what it is representing with its CIPs? The representation must include more than just the self; which is to say, my CIP avatar becomes aware of the world it encounters because it either constructs other CIPs to represent worldly experience within its own circuits or else it can "read" the CIP representations that otherwise would remain unconscious (see figure 4.3).

Maybe we should frame the issue this way: How are we consciously aware of our sense of self? It is the self that engages the world. If our avatar is conscious of itself, then it is automatically aware of the self's operations. This is equivalent to asking how the conscious-mind avatar is aware, at least in some limited sense, that it is an avatar. Perhaps the avatar is self-aware because it was created by the brain to represent the self, not to represent the sensory world, but to serve as an interface to it. Both unconscious mind and the conscious avatar interface with the real outer world and with each other.

The avatar CIPs, representing the sense of "I," are accessible to the unconscious mind's operations, which generate the avatar. That is, the brain knows that it has this avatar and knows what the avatar is doing, even if only unconsciously. The avatar, however, knows consciously because its information is processed differently within the context of the sense of "I." Stimuli are not isolated physical events. The *representation* of events is self-conscious awareness.

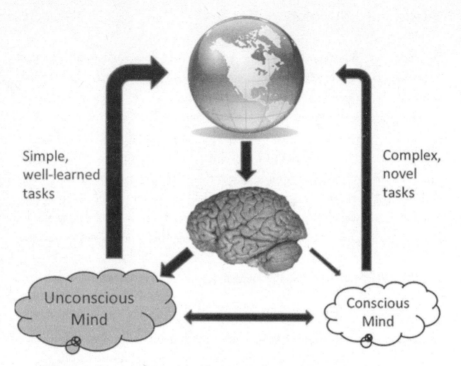

Figure 4.3. Summary diagram of the avatar function. The unconscious mind and the avatar (conscious mind) interact with the external world and with each other. Discrepancies in the size of the various arrows is intentional, reflecting the predominant influence of unconscious mind.

The avatar also registers sensory input and its own self-generated intent, decisions, plans, and the like in its own terms—all referenced to itself. Thus, it achieves these registrations consciously.

CIPs contain the representation of unconscious thought. Thus, a first step in enabling conscious awareness of such thought would be for the various circuits to have overlapping elements; that is, certain neurons would participate in several or more circuits and in that sense monitor the information contained therein and share it among all shared circuits. This is a physical way by which circuits can be made "aware" of what is going on in other circuits.

Such monitoring need not be continuous. Neurons participating in shared circuitry could be gated, through inhibitory processes, to sample activity in other circuits by a process akin to multiplexing in an analog-to-digital converter that sequentially samples activity one data channel at a time in multiple signals. That process would allow the brain to sample across multiple streams of its unconscious thought in parallel pathways, resulting in a combinatorial code. Multiplexing across parallel circuits provides a possible mechanism for the "binding" phenomenon.

Multiplexing may seem counterintuitive, especially since we think that humans multitask. Young people seem especially able to do multiple things at the same time, such as text messaging on a cell phone, driving a car, listening to a CD, playing a video game, and talking to a friend. But this is an illusion. The brain really does only one thing at a time. Our brains work hard to fool us into thinking that it can do more than one thing at a time. Recent MRI studies prove that the brain is not built for good multitasking.[20] When trying to do two things at once, the brain temporarily shuts down one task while trying to do the other. In that study, even doing something as simple as pressing a button when an image is flashed causes a delay in brain operation. The MRI images showed that a central bottleneck occurred when subjects were trying to do two things at once, such as pressing the appropriate computer key in response to hearing one of eight possible sounds and uttering an appropriate verbal response when seeing images. Activity in the brain that was associated with each task was prioritized, showing up first in one area and then in the other—not in both areas simultaneously. In other words, the brain only worked on one task at a time, postponing the second task and deceiving the subjects into thinking they were working on both tasks simultaneously. The delay between switching functions can be as long as a second. It is highly likely, though not yet studied, that the delays and confusion magnify with increases in the number of different things one tries to do simultaneously.

Could we be consciously aware of our other senses, of smell, taste, sight, hearing, and so on, without having a sense of self? Not likely. The unconscious mind can register these sensations, but not consciously. In the real time during which unconscious mind registers sensations, the consciousness avatar must be perceiving the sensations. How can this be? Is there some shared access to sensation? How is the sharing accomplished?

Consider the following example in which the eyes detect a tree. The image is mapped in unconscious mind by a CIP representation of the tree. The brain searches its circuits for a template match in memory. When a match occurs, the brain searches further in memory stores for other associations, such as the word *tree* and any emotional associations. The memory CIPs are then accessed by the CIPs of the consciousness avatar, which becomes aware of what the unconscious mind has processed (and does so in its context of self). The avatar says, "I see the tree."

Conscious mind monitors and adjusts its representation of itself as necessary. It also monitors some of the CIP representations of unconscious mind, but it presumably has no direct access to the operations of unconscious mind. The representations of self in conscious mind can do other freewill kinds of things, such as reflect on what it knows, plan, decide, and veto. In other words, the conscious mind is a "mind of its own."

The Avatar's Artifacts

Avatar power is enhanced by the "tools" it creates and uses. These include gestures and body language, written and spoken language, art, music, and physical tools (hammers, computers, and so forth).

Remember, the function of the avatar is to serve the interests of its embodied brain. A major way the avatar does this is through the construction of artifacts, objects or processes that the brain can find useful in some way. This includes most obviously tools. All such artifacts reflect the avatar's desire to engage itself, and in turn its brain, in the ways of the world. To the extent possible, the avatar also wishes to manipulate the ways of the world to its purposes.

Note the use of the word "purpose." Avatars create their own purposes. Purpose becomes explicit in consciousness and is a cardinal feature of biological avatars. The purpose and utility of good artifacts is that they extend reality in creative and practical ways. One has to wonder if this same imperative operates in the brain's creation of its avatar. That is, the brain may deliberately create an avatar that is an embellishment of the *real* unconscious self it is trying to engage with the world. Unconscious mind needs tools too, and the best way to get them is to generate an avatar.

Beliefs. Avatars construct belief systems, which then serve as frames of reference for interacting with the world. Though Andrew Newberg and Mark Waldman did not think in terms of avatars, their book made a compelling case that beliefs are biologically based.[21] They even say that humans have an inherent "will" to believe. Though beliefs are not typically thought of as mental tools, they most certainly are just that. Ultimately, humans always turn to their beliefs and use them as tools to guide their interactions in the world.

Gestures and Body Language. This form of externalizing thought is typically regarded as a way to benefit the thinking of the one who watches somebody else doing the gesturing and body language. But there is evidence that such movements in space also help the thinking of the person trying to communicate. For example, if I know something is very small, my thinking about it is augmented by the movement feedback I get as I hold my hand out and bring the index finger near to the thumb to show people my impression of its size. If I want to indicate the size of the fish I caught, it is a lot easier to hold my palms sideways and spread my hands apart by the estimated distance.

Gesturing can make abstract ideas more explicit. Moreover, because the idea is expressed by a display in space, it becomes more memorable, both to the observer and to the gesturer. This is because the brain's spatial mapping system, in the hippocampus and entorhinal cortex, is also the system that converts temporary memories into lasting ones. Thus, spatial displays provide cues needed by the spatial-mapping systems to help form the memory of the displayed information.

I once attended a seminar by psychologist Barbara Tversky on this topic. A member of the audience asked about how someone who can't move, like Stephen Hawking, the famous physicist who has ALS, thinks so effectively when he can't create externalized thinking tools. My answer to the audience was that the premise of the question was wrong. Hawking does have an externalized thinking tool. It is called mathematics, which he surely uses in his "mind's eye blackboard" for displaying equations and working out solutions. Professional athletes do a similar thing by visualizing themselves in their mind's eye performing a perfect golf swing, pole vault, or other action required in their particular sport.

So the point is that the avatar has an amazing ability to create externalized thinking tools, either through having them created by the body or by imagining them in the "mind's eye."

There may be a "freewill" implication here. The avatar can choose what kinds of externalized tools it wants and how it wants to use them. If I want to create a physical thinking tool, like a map, for example, I can direct the creation with as much or as little detail as the avatar wants. Nothing internally compels an avatar to use a specified amount of detail. If the avatar wants to practice a golf swing in the imagination, for example, it can choose between a complete swing, a backswing, a follow through, a putter movement, or whatever. In such cases, the CIPs of the avatar get to decide what to do. None of the options are mandatory.

Art. As human culture evolved, it was initially rendered in drawings and carvings in durable material, such as stone. From crude ancient drawings on the walls of caves in France, we have concluded that even primitive humans had rudiments of a conscious avatar that wanted to display its own representations of the world. These primitive avatars clearly wanted to represent themselves and their ideas, but they also had a notion of posterity. They probably wanted their artifacts to be a historical record to be seen by the avatars that would succeed them.

Music. Music is another fundamental artifact used by avatars for communicating. Music, like art, is a language of emotion. Emotions are the province of the unconscious mind, yet only a conscious avatar can create music in its most robust forms.

What is it an avatar is trying to express via music? Most likely, music is created by the avatar to express emotions and even ideas that the avatar cannot find words to express. Music is the avatar's way of showing how it feels or wants others to feel. Of course music is not an animate thing, but it does serve the purpose of changing how we feel. It also is a stimulus that satisfies the cravings of our own avatar to seek stimulus. So here we have the case of the avatar creating music to affect others but also to satisfy its own needs.

Tools. Even the modest brains of certain animals have the capacity to create rudimentary tools. And it is not necessary for the animal to have a conscious avatar. For example, on his trip to the Galapagos Islands, Charles Darwin saw a species of finches that used twigs to probe for insects. We have known for a long time that chimpanzees use stones to break things like nuts. Sea otters use stones to break mussel shells. In 1960, Jane Goodall made a splash in the behavioral ecology world by reporting that chimpanzees dig into termite mounds with a large stick and then draw termites out with fishing-like motions. Others have reported that some chimps use spears that they sharpen with their teeth and then use to kill bush babies.

Traditionally, tool use has been regarded as an indicator of conscious intelligence. There is the classic experiment involving suspending bananas out of reach from the ceiling of an empty room that contains only a chimp, a box, and a stick. Chimps quickly figure out they can get the bananas by grabbing the stick, standing on the box, and using the stick to knock down the bananas. Other species can't do that (of course, most other species don't have hands they can use to grab the stick). Anyway, the interpretation is that chimps are smart. One could just as well conclude that they have a conscious avatar that is serving the interests of its embodied brain through conscious intent, thought, and planning.

As human civilization has advanced, so has the scope and usefulness of human tools. Our own era is dominated by tools that make and operate things, such as cars, planes, home appliances, air conditioning, all sorts of electronic technology, and the manufacturing tools needed to make such artifacts.

To date, invention of the computer is the ultimate artifact creation. Computers duplicate many of the actions of human avatars and are superior in many ways (such as playing chess). Nonetheless, let us remember where computers come from. They are the creation of the human avatar.

Humans do something very crucial that no computer or other animal can do as well: the human brain externalizes its cognitive processes. This may well be the avatar's way to manipulate and expedite its information processing. Such tools are created and used consciously, and that is why I suggest these tools are the special province of the avatar.

The avatar also uses certain artifacts in a holistic and superadditive way (as in reading a section of a map or in "bigger-picture" terms of perceiving the artifact). Such processing by the avatar is known as *gestalt*. It is a very useful cognitive tool that enables the avatar to perceive extra dimensions to the artifact, as in "the whole is greater than the sum of its parts." Such ability underlies the capacity for perceiving artifacts in their totality and for creative synthesis and invention.

What about situations where externalization is not physically feasible, but can be simulated, in the "mind's eye" for example? We can externalize things like maps, art, equations, and the like in our mind's eye and use the virtual representation as a thinking tool. There is also something like a "mind's ear," wherein musicians at least can think about the sound of music without actual physical sound.

Virtual representations are much less common for other senses. Can you image a "mind's nose," or "mind's taste," or "mind's touch"? These sensations are much less imagined because not many neurons are dedicated to these senses. The sensory cortex, for example, has only a small area set aside for taste and touch. The sense of smell is not directly represented at all in the neocortex.

Language. Like tool use, certain animals have the capability of sound communication. Songbirds, whales, nonhuman primates, and a variety of other species have a repertoire of sound that they use for communication. But sound communication that goes beyond a stereotyped repertoire might require the mediation of a conscious avatar. In humans, language is clearly the most powerful artifact.

Religion. Religion is both an individual and collective avatar artifact. Throughout all history, human avatars have constructed various religions as a tool to create, validate, and enforce codes of beliefs, ethics, and behavior. Thus, religion is a tool that makes society less chaotic and more governable. Religion is also a tool to salve the emotional trauma brought on by thinking that death is the end.

Biologist E. O. Wilson asserts that most religions are like organisms: they "have a life cycle. They are born, they grow, they compete, they reproduce, and, in the fullness of time, most die."[22] At the core of each religion is a creation myth, which often conflicts with the findings of science.

Religions have "tribal roots," which produce corresponding discrepancies in belief systems in different communities. Some belief systems are compatible with science, but they can't all be compatible because they differ among themselves. Religious doctrine grows out of theories, just as scientific doctrines grow out of theories, which, however, are derived from evidence rather than from faith. However, remember what I said earlier about scientific theories. They are accepted as a matter of faith until there is conflicting evidence.

Avatars Learn and Teach

There is no known hard wiring that we know of for consciousness. Unlike the traditional five senses, the brain must *learn* how to construct a neural representation of a conscious sense of self (SoS). The brain must learn what circuits to recruit and what CIPs and spatio-temporal codes to use to represent a conscious self. Presumably, this process becomes more and more automatic as our brain follows its destiny from womb to tomb.

The brain's avatar representation is constructed in long-term memory over the years. Yes, the sense of self is learned. The only things that change that representation are incremental learning, which produces changes to the SoS as one ages, and the complete destruction of SoS that occurs with Alzheimer's disease.

Note that the sense of identity comes back after sleep and changes only slightly from day to day. That is, the CIPs that represent our sense of self are stored in long-term memory and always come back intact into realization when consciousness is enabled by the requisite brainstem-cortical interactions.

Our avatars learn from engaging the world. The five bodily senses provide our "I" with rich experiences, many of which we use and remember for using in subsequent experiences. Our avatars become what they are because they have significant control over what they choose to experience. They seek out certain kinds of stimuli and experiences. They reject some learning experiences and embrace others. They recognize things they do not know and contrive ways to increase their knowledge and understanding.

As a CIP representation, the avatar has the same opportunity to learn and be modified by experience as do the CIPs for unconscious sensory and motor operations. When I was a baby, my brain was enriching its topographical maps and teaching me who I was and what my place was in the world of mother's milk, dirty diapers, and the crib that restrained me. As life progressed, my new experiences expanded my sense of self. No doubt this was made possible by the creation of the many new circuits that form in babies as neurons grow their fiber branches and create more synaptic contacts.

And if I should be unlucky enough to develop Alzheimer's disease, the CIPs supporting my conscious SoS will shrivel to the point where the brain no longer generates a complete avatar. I will eventually lose what I know about myself, and even my sense of self altogether, even though I may still be able to move about and experience somatosensory sensations.

If the brain can generate its own avatar, you would think the brain would make the most useful and worthy avatar possible. Why then are the avatars of so many people less than flattering? The answer lies partly in the fact that the avatar is only one facet of brain operation. The unconscious brain also generates compulsions, emotions, and irrational behaviors that often conflict with whatever the avatar might wish to do. The other part of the answer is that the brain does not always know what is in its best interests. Much of this has to be learned, and that takes time. Kids are likely to do foolish things. Wisdom grows with age. Much of this learning is accomplished through the conscious learning of the avatar, which constrains the actions of the demons and fools of unconscious mind. In other words, unconscious mind relies on its avatar to teach it how to become the best it can be.

Collectively, our avatars teach each other, sharing what individual avatars have learned and in the process creating group culture. It is our avatars that have taught us about life force, particularly that of our own. In the formal collective sense of science, to accomplish the understanding we have to date, our avatars had to create an extensive array of tools, such as telescopes, microscopes, electronic instruments. This reminds us of yet another distinction between living and nonliving things. Life, at least higher forms of life, creates avatars, and those avatars create artifacts that they use to enrich their engagement with the world.

ARE WE FREE OR ARE WE ROBOTS?

A hallmark of being human is the capacity for generating beliefs, intentions, and presumably free decisions. There is no doubt that humans exhibit willful, considered behavior. We have, many people say, free will. That is, we supposedly can decide from among several alternatives, and this process is not directly controlled by external or internal imperatives. The human avatar says, "I not only think, I also am in charge of what I think and do." As we shall see, this view is widely viewed as wrong. It is fashionable among academics to assert that free will is an illusion. I disagree.

In a conscious state, the brain's global workspace expresses its CIPs in the form of the conscious avatar. The avatar operates within the workspace as an actualized and partially autonomous being. As such, these CIPs can process information and choice options in a state of explicit self-awareness. Because the avatar operates in such a mode, it creates beliefs and choices that are consonant with the nature of its being, which includes established biases, values, beliefs, goals, intentions, and plans. Because the avatar understands things with explicit awareness, it has a degree of freedom in generating beliefs, intentions, choices, and plans that is not available to unconscious mind. Skeptics will say that these aspects of the avatar's nature were preprogrammed by past experience or by genetic disposition. I say that experiential influences are developed and learned and become explicitly incorporated because they are acceptable to the avatar. Acceptance or rejection can be largely a free choice. Even much of the genetic influence is actually epigenetic, in which a person's choice of thought, behavior, and experiences helps to control which genes get expressed.

How Choices/Decisions Are Made

Nobody knows for certain how decisions are made except at the level of nonconscious mind. In simple reflex circuits for example, whether or not an input generates a reflex output depends on accumulation of excitatory influence in the chain of neurons that mediate the reflex. If excitation reaches threshold, then output results. It seems reasonable to apply this principle to higher brain functions that result from pools of neurons that control a given state.

"Decision neuroscience" is an emerging field of study aimed at learning how people make decisions and optimize the process. Neuroscientists seem to have homed in on two brain-level theories, both of which deal with how the brain handles the processing of alternate options to arrive at a decision. One theory is that each option is processed in its own competing pool of neurons. As processing evolves, the activity in each pool builds up and down as each pool competes for dominance. At some point, activity builds up in one of the pools to reach a threshold, in winner-take-all fashion, which allows the activity in that pool to dominate and issue the appropriate decision commands to the parts of the brain needed for execution.

Such a process could explain the state changes that occur when you mull over a decision. If you are trying to decide on whether to have fish, chicken, or beef for dinner, your brain processes these options in competing pools of neuronal circuits to yield eventually a preferred option. As activity builds up in these competing populations, one pool of neurons gradually prevails, and its choice becomes the winner.[23]

In the example above, the null state is the "choice not made." As the winning choice emerges, a sudden transition occurs, where the contest is over and the choice is made (fish, for example). This notion of sudden transition is fundamental to brain operations of many kinds. The most profound transition is the sudden falling asleep from a state of wakefulness. The reverse also applies. We suddenly transition from sleep to wakefulness. In the morning, many of us wake up in a seeming instant, as if a light switch turned consciousness on. For others, the transition is more like a computer booting up.

The other theory is based on guided gating. Here the idea is that alternative option processing controls the flow of activity in a network of brain areas to direct the flow of information of one option toward the appropriate parts of the brain needed for execution. Input distributes in parallel to pools of decision-making neurons and is guided to regulate how much excitatory influence can accumulate in each given pool.[24] The specific guided gating would involve inhibitory neurons that shut down certain pathways, thus preferentially routing input to a preferred accumulating circuit. The route is biased by past learning, current emotional state, and the ongoing situation and contingencies.

All of these decision-making possibilities involve a concept called *integrate and fire*. That is, input to all relevant pools of neurons accumulates and leads to various levels of firing in each pool. The pool firing the most is most likely to dominate the output; which is to say, the decision.

Regardless of how circuits make decisions, there is some evidence that CIP representations for each given option simultaneously code for expected outcome and reward value. These value estimates update on the fly.[25] Networks containing these representations accommodate such information to arrive at a decision.

The issue at hand is whether decision making or choice is made freely. The mechanisms just described seem amenable to manipulation by avatar operation.

Both conscious and unconscious choices are no doubt modulated by stored memories of past experiences and the values the brain has placed on them. Conscious mind brings these stored memories and propensities into working memory, whereupon the relevant factoids, values, propensities, and so forth are streamed into the explicit thought of the avatar. The avatar, being self-conscious, presumably has a greater degree of freedom than does the unconscious mind in making decisions because it is explicitly aware of the presence of and conflicts among the various options that are available.

Consciousness seems to lie at the heart of issues about free will and personal responsibility. Unconscious mind can will actions, but it makes no sense that such will could be free. Conscious mind is not aware of the processes of unconscious activity, such as pools of competing neurons or guided gating, but it is aware of the consequences of such activity. In theory, no longer is the conscious brain limited to passive execution of existing programs launched from the unconscious mind; conscious mind might allow a more comprehensive consideration of what is being experienced. More information is accessible and coordinated. More options can be analyzed in the daylight of conscious working memory, and this analysis can be more comprehensive. The avatar of conscious mind can make the brain and body do things. Voluntary choices become available. New programming is facilitated.

There are those who believe that conscious mind is not in charge of what it thinks and does, that conscious mind is only a mirror that shows us the results of some unconscious processes. The avatar is regarded by many scholars as an observer, not a doer. They say free will is an illusion.

This notion of illusory free will dates back to ancient times. Among the first to make the point was Buddha in the sixth century BCE. While there is much to recommend in Buddhism, which is more of a lifestyle philosophy than a religion, the teachings of this philosophy are confused about the existence of self and free will. Buddha held that human mind exists but that the "I" of self and ego do not exist. There is no soul either. Indeed, belief in self is considered the cause of all the evil in the world. Because everything is interdependent and relative in a natural world, free will cannot exist.

Buddha taught that one achieves serenity and happiness by being aware of what the mind urges us to do and how it reacts to inevitable vicissitudes of life. Buddha instructed us to be aware of our wants, angst, anger, and feelings in general so that we may discipline the mind to serve our best interests. Note the contradiction. How can you do that if there is no "I" capable of introspection, watching the mind, striving to be aware of bad thoughts, and supplying corrective discipline? Comedian Tim Allen, in his semiserious book *I'm Not Really Here*, challenges Buddhist philosophy this way: "Eastern mystics say that there's no one home inside us, no 'I' even asking the question. The 'I,' they claim, is an illusion. . . . So tell me smart guys: who's doing the looking?"[26]

In fairness to Buddha, he had no way of knowing that the brain creates its sense of self from the neural body maps formed in the womb. Nor could he have known that consciousness might exist as CIP representations of an avatar that can make brain and body do things.

Among scientists, the view of illusory free will was originally espoused by some of the most prominent scientists in history, including Thomas Huxley, Charles Darwin, and Albert Einstein. A growing number of contemporary neuroscientists agree.

In our time, the "movement" received a big push from the advent of behaviorism, originated by John Watson and enshrined through the research of B. F. Skinner shortly after World War II. Behaviorism was a philosophy that regarded human behavior as a "black box" of environmentally conditioned actions. Conscious thought was not considered appropriate for scientists to think about or study. Fortunately, a significant segment of the scholarly community has rebelled against behaviorism and created in its wake the new field of cognitive neuroscience. In this field, study of consciousness is not taboo; it is necessary.

Addictions are often cited as proof that humans don't have free will. Addictions develop over time, and thus constitute a learning process for the brain. The addicted brain has been reprogrammed. Clearly, at least with regard to the addicted brain, there is little capacity for free will. But there once was. Former cigarette smokers believe they deserve credit for a freewill choice to quit. I do, knowing firsthand the magnitude of conscious force of will it takes to quit that terrible habit.

For semantic convenience, I will refer to the illusory-freewill people as *robotocists*, as their position requires the assumption that humans are biologically programmed robots. One reason for the robotocist perspective is that our brains have been programmed by our genetic endowment and by past experience—

thus, the conclusion follows that we are biological robots. The probability is that any given action cannot be freely controlled because we have built-in biases and propensities that unconsciously control the generation of intentions, choices, and decisions.

Before we explore my position, perhaps it is useful to consider the history of how scholars have thought animals make choices and act upon them. One of the more obvious actions is movement, which is typically considered as either "voluntary" or "involuntary." An involuntary movement might be, for example, an animal's escape behavior when it is being chased by a predator. A voluntary movement might be one where an animal moves when it did not have a compelling need to do so, as my dog does when she moves from one sleeping spot to another. Such a voluntary movement entails a choice, and since it occurred when the animal did not need to move, you might say that it was willed. How "free" that willed action was is what is at issue here.

Some surprisingly relevant studies have recently been conducted in the primitive brain of crawfish.[27] Crawfish will often move when they don't have to. This movement is preceded by increased nerve impulse activity in restricted parts of their brain. It is therefore called a readiness discharge. The impulses serve to prime activity in clusters of neurons in the body that are pattern generators for walking. Clearly, the crayfish does not consciously decide to walk, nor does it decide how to do it. Even so, such an action can and does occur when it does not have to. No external stimulus is required. There are, however, internally generated stimuli that turn movement-initiating systems on. Likely, such movement results from random neural activity in the pattern generator circuit, and the option to remain in place or move to a different place was always there. Random things happen in nervous systems. They are not always inevitable. Does that make them "free" in some sense?

These internally generated stimuli may be something as simple as inherent membrane instabilities in certain movement-initiating neurons that make the neurons discharge impulses. If that is the case, it is misleading to use the word "voluntary" to describe the movement. It would be even more misleading to say this behavior arose from free will, given that the crayfish brain is too primitive to have such capability.

The neurons that generate this readiness discharge produce a sequence of excitation and inhibition in neuronal populations that do not depend on feedback signals from downstream targets in the body. The issue is whether these readiness neurons are triggered from upstream neurons or are self-generating within their own circuitry. The readiness discharge does have a purpose: it expedites the signal processing for self-generated initiation of movement.

Higher animals, including humans, have comparable readiness-response neurons associated with movement. In humans, increased activity of readiness neurons in part of the movement-control portion of the cerebral cortex has been seen to apparently precede a conscious decision to move (see below).

The Biological Robot Argument

Most nonacademics tend to take as a given that people have intentions and can freely make decisions and choices when there are alternatives and an absence of external constraints. To regular people, free will is self-evident. We think of free will as follows: when faced with two or more alternatives, we have the ability to consciously and freely choose one of the alternatives, either on the spot or after some deliberation.

But sophistry prevails among a growing number of scientists and philosophers. They view humans as biological robots that can only do what their nonconscious and unconscious minds tell them to do. Robotocists arrive at their counterintuitive conclusion by stressing that our brains are made of molecules. Molecules (and the neurons they make up) must obey the deterministic laws of physics. Our "decisions" are made by brains, which are in turn made of molecules. Therefore our brains must obey the laws of physics. Absent any quantum indeterminacy, what our brains do is determined and predicted by the laws of physics. Ergo, in any situation we cannot make a free choice. Physics does it for us.

Robotocists buttress their position using speciously interpreted research that does seem to challenge the traditional commonsense view of free will. Let us consider their evidence against free will.

People with brain injuries provided the first arguments against free will. For example, people with injuries that caused amnesia were studied by British psychologists Elizabeth Warrington and Lawrence Weiskrantz.[28] They showed a series of words to the amnesiacs, who could not remember the words. Then the patients were shown the first three letters of each word and asked to complete the letters to make a word, any word. Amazingly, they consistently conjured a word that was exactly the same as the one they had just seen and forgotten. In other words, the words had been memorized in the unconscious mind but not in the conscious mind. The conclusion was that word selection was made unconsciously. There's another explanation the researchers did not consider: inherent memory-recall deficiencies. Conscious recall is often difficult, and it requires multiple cues, which the experimenters provided. Even normal people frequently have trouble recalling information that they know well.

The robotocist argument may have begun catching on with the 1976 book *The Origin of Consciousness and the Breakdown of the Bicameral Mind* by Julian Jaynes.[29] Jaynes gave many logical arguments that consciousness is not necessary for thinking and that most human mental work is done unconsciously, only becoming realized consciously after the fact. Jaynes concluded that consciousness is used only to prepare for thought and to perceive and analyze the end result of thinking. He did not explain why brains evolved this capability when they can't do anything about their conscious analysis.

Subsequent theorists argue that decisions and intentions are made unconsciously and that the egotistic conscious mind then lays claim to them as its own (see figure 4.4). This position holds that the brain is an automaton that creates its own rules and makes sure that we live by them. There is no "I" in charge. The brain is in charge of itself.

But think about what was just said. A brain that is in charge of itself could still do so from consciously driven intentions and choices. Your avatar, for example, could tell your brain to make you stop smoking, or eat less fat, or read certain uplifting books.

Robotocists cite the existence of compulsions and addictions as examples of conscious awareness failing to control the brain. In such cases, conscious mind knows we have bad behaviors but can't do anything about it. Our excuse is that we are addicted, have a brain disorder, or have been programmed by bad events beyond our control.

The same kind of logic is used to explain character or personality flaws. How convenient! People say, for example, "I can't believe he did that. He is such a good boy" or "he can't help it, that's just the way he is" or "she really doesn't mean to be that way." Robotocism is the mother of all excuses.

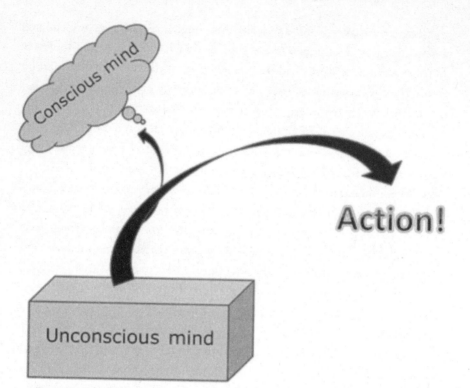

Figure 4.4. The concept of illusory free will. Unconscious mind is said to create behavior and belatedly inform the conscious mind of what has already been done.

A more formal philosophical argument is provided by Henrik Walter.[30] He says our standard theory of mind is a mere convenience that satisfies our expectations about what we and others do. Walter says that criminals cannot be held responsible for their crimes. They may be self-aware and even regretful about what they did, but they could not stop themselves. He emphasizes that the conscious mind can only "look in" on what the unconscious mind is doing. In other words, our avatars can see and hear but not do. A more formal elaboration is that free will operates "to ensure the continuity of subjective experience across actions which are—of necessity—executed automatically."[31] Can somebody tell me what the point or value is of ensuring continuity of subjective experience if nothing comes of it?

A complete defense of the robotocist school of thought can be found in

Daniel Wegner's *The Illusion of Conscious Will.*[32] Leading thinkers, such as the philosopher Patricia Churchland[33] and the neuroscientist Michael Gazzaniga,[34] recognize the nihilistic nature of the robotocist conclusion but are resigned to a position of "it must be so." The most recent book on this matter perpetuates the robotocist argument at least for many short-term intentions and asserts that the question remains open for all other intentions.[35]

Philosophers seem to polarize around two points of view: (1) that people generally lack free will but sometimes may have it (compatibilism) or (2) that human thoughts are beyond personal control and incompatible with free will (incompatibilist). Incompatibilists argue that no empirical tests have been devised to prove that free will exists and, moreover, that experimentation proves their view that free will is an illusion. They uncritically accept research that purports to prove free will is an illusion.

Robotocist Research and Its Critique

In modern times, the freewill conundrum has been exacerbated by neuroscience experiments that many scholars accept as proof for robotocism. The accumulation of evidence began with simple experiments performed and elaborated in the 1980s by the deceased University of California scientist Benjamin Libet. I visited with Libet at a brain-research conference in California, but this was before his famous experiments, and I never got to discuss that research with him.

My analysis will focus on the prototypical Libet experiment and those of others that followed. I recently published the essence of this critique in a peer-reviewed journal.[36] The critique examines the two main lines of research that provided the scientific underpinnings for modern robotocism. One is the paradigm developed by Wegner in the 1990s in which subjects were asked to move a cursor randomly around a computer screen and stop the cursor every thirty seconds or so over an object depicted on the screen. The system was programmed so that some of the time the computer stopped the cursor. After each stop, the subjects rated their intentionality in terms of how sure they were either that they had made a conscious decision to stop the cursor or that the experimenter had done the manipulation behind the scenes. In turns out that subjects were quite bad in making such estimations. They were correct in believing that they had actually caused all of the stops only 56 percent of the time (pure chance level was 50 percent). Wegner developed a later approach by having subjects view other people's gloved hands located in the position where their own hands would be. As the gloved hands performed actions, subjects were asked to rate the extent to which they had controlled the movements. Again, subjects performed poorly in such estimates.

I reject his robotocist conclusion because his experimental designs could have been measuring awareness as well as intent. The major uncontrolled variable is the level of reliability of the subject's awareness of his or her conscious intent. Also, one cannot conclude unequivocally that the intent is either conscious or unconscious. As we shall see later in my comments on Ben Libet's experiments, awareness of conscious intent is quite problematic. Tim Bayne has written a more exhaustive criticism based on the difficulty of being aware of intent.[37] In any case, intent and awareness of intent are two different things.

The Libet experiments. Ben Libet's experiments have been more widely accepted at face value. Libet monitored a "voluntary" finger movement while at the same time recording the EEG from the scalp overlying the part of the cortex that issues movement commands to the fingers.[38] Participants were asked to make a spontaneous finger movement at a time of their choice while watching an electronic spot moving around a clock face. Subjects were to observe and remember the time on the clock at the instant they decided to move the finger. When subjects consciously decided to make a movement, they reported the time of the decision from the modified clock. As expected, subjects thought that they had decided to move about a half second before actual movement, which is consistent with the idea that they willed the movement to occur. But the startling finding was that a major change in neural activity in the premotor cortex was observed about a third of a second before the subjects claimed to have willed the command to move. The interpretation of such a result is that the decision was made unconsciously and that conscious awareness realized it after the fact. Accepting that premise, one is forced to conclude that one does not consciously "will" such movement. The brain unconsciously decides to move the finger and then lets the conscious mind know what it has decided (as illustrated in the above figure suggesting free will is an illusion). The disturbing corollary is that one does not freely "choose" to do anything. The brain is just driven by external and internal forces to direct behavior, and one's conscious avatar is only around to know about it.

Follow-up studies. In a similarly designed follow-up to the Libet experiment performed by H. C. Lau and colleagues at Oxford, human brain scans were made as subjects were asked to report when they first generated the urge or intention to move.[39] The images showed three small neocortical regions of activation when the subjects attended to the urge to move prior to the actual movement itself.

The Lau studies showed that subjects reported an intention to move about a quarter second before the actual movement, which is consistent with Libet's results. Conscious intention was associated with increased neural activity in areas other than the motor cortex. These activations might have occurred before the motor cortex was activated, but the imaging method used did not have the time resolution to answer this question.

Nonetheless, these results showed other experimental limitations. This is reinforced by the findings of Sukhvinder Obhi and Patrick Haggard,[40] who found that awareness of conscious intent correlated more specifically with a motor-cortex potential over the side of the head *opposite* to the hand making the movement (hand movements are initiated from the opposite cerebral hemisphere). The motor cortex potential occurs in the motor cortex of both hemispheres, but the conscious awareness only correlates with the side that controls the hand movement. Thus, the lateralized motor-cortex response indicates how the action will be implemented (by the hand) rather than general conscious intent.

A follow-up study by the Lau group did examine more closely the timing judgment issue.[41] Specifically, the group examined Libet's finding that subjects misestimated the onset of movement, thinking it occurred about 50 msecs (or five hundredths of a second) before it actually did. The Lau group reasoned that there must be someplace in the brain that signals the judgment that movement has occurred and that across subjects the magnitude of the brain-activity correlate would positively correspond to the accuracy of the time estimate. Alternatively, enhanced electrical activity might contribute to the time-estimate error, in which case the correlation would be negative. They also re-examined their earlier brain scan data to see if the same principle applies for judgment of the onset of intentions.

What they found confirmed many earlier studies that indicated that the brain makes errors in time estimation. When participants were required to estimate the onset time of their movements (instead of their intentions), the brain activity was enhanced in part of the motor cortex. Moreover, across subjects the level of activity was positively correlated with time-estimate accuracy. That is, the greater the activity, the better subjects estimated the time of the movement. This is consistent with the group's earlier data on brain-scan changes in a motor area and time estimates of onset of intention. In both cases, time estimation could not be relied upon as accurate.

The recent studies by Chun Soon and colleagues also used a brain imaging design like that of Libet's experiments.[42] They monitored brain activity in the same area as did Libet, the supplemental motor area (SMA) of the cortex. However, they reasoned that the SMA is active in the late stages of a movement decision and that other brain areas might be involved in movement planning at earlier times. They also used a more sensitive way to establish when the awareness of a decision occurred. Finally, they gave subjects a choice between two buttons to press.

What Soon and colleagues found was astonishing. Two regions in the frontal and parietal cortex exhibited a decision-predictive change a full seven to ten seconds before conscious awareness of the decision. The areas of the motor cortex that actually issue movement commands showed slightly increased activity in the second or so immediately prior to the instant of the decision, and they showed much more pronounced activity after the decision. Activity in brain areas directly involved in issuing movement commands (the SMA and the motor cortex) increased greatly after the decision.

The increased activity in the other areas prior to awareness can be interpreted in more than one way, though the authors were wedded to just one interpretation. Most people, especially the lay press, assume that these other areas are processing the decision to move and thus indicate absence of free will because their increased activity occurred before subjects thought they willed a movement. However, the authors were careful in wording their conclusion;

namely, they stated that the frontal and parietal cortical areas "influenced" the decision making up to ten seconds before a conscious decision to press one of the two buttons was realized. They viewed this early, predecision activity as preparatory and also as a specific predictor of which button was to be pressed.

As with the original Libet experiments, experimenters relied on self-report of the decision to move, which has limited time resolution. Also not considered was that these experiments had the same limitation in that one must presuppose that the decision to move as well as the conscious realization is instantaneous. Both assumptions are wrong, as I will explain later.

A replication of the Soon study used electrical recordings.[43] Not surprisingly, electrical changes from multiple scalp electrode locations occurred several seconds before subjects indicated a conscious decision to move. These results were again interpreted as indicating that such antecedent activity reflected unconscious decision making, as if that is the only logical possibility. I claim that the interpretative flaw remains: decision making is *assumed* to be unconscious. But where is the actual evidence for that? All such data really prove is that there is antecedent neural activity, some of which almost certainly involves conscious working memory of the rules of the game.

In designs like this, the subject continuously and consciously processes what is supposed to happen in the experiment. He or she knows as soon as one trial is over that another is beginning. Moreover, the subject consciously knows to choose to make a movement. The brain no doubt is planning to make such a movement long before a "go" signal is delivered via the conscious decision-making process. So, the premovement increased brain activity could actually reflect conscious processing in working memory of the rules of the game that the subjects have willfully agreed to obey. All through a trial, before conscious decisions are made, the brain is consciously processing in working memory at least six different things: (1) "I will make a button-press movement," (2) "I will press either on the right or the left," (3) "I will notice the letters on the screen and hold them in working memory," (4) "I will issue a go decision when I feel the urge," (5) "I will remember which letter was present on the screen when the go command is issued," and (6) "I will have to select that letter from a screen display of several other letters with yet another button press." Under these cognitive conditions, it is unrealistic to expect any single brain-function metric to mark when a decision is made. Yet, even so, this study actually supports a nonrobotocist interpretation. Activity did increase in nonmotor, consciousness-mediating areas before the movement. In other words, the "go" decision was only one part of the consciously directed process.

A refreshingly different approach was that of Michel Desmurget and colleagues in France.[44] First, they distinguished between two processes, the will to make movements and the awareness of such willed action. Second, they used direct electrical stimulation rather than recording. The subjects were conscious humans with electrodes inserted into the brain to help locate tumors that turned out to be located elsewhere. Stimulating the right inferior parietal regions triggered a strong conscious intention to move the hand, arm, or foot on the opposite side of the body whereas stimulating the left inferior parietal region produced an intention to make the movements of speaking (not surprising because this area includes a speech center). When stimulation strength was increased, subjects believed they had actually made such movements, even when monitoring of the relevant muscles showed no signs of muscle activation. This result does not fit the robotocist theory because there was a clear sign of consciously willed action even when no movement occurred. Other researchers had shown that low-intensity electrical stimulation of the supplemental motor cortex in humans caused an urge to move without accompanying movement. Stronger stimulation caused actual movement.

The lay press has claimed this is proof of free will. I won't go that far because the data simply show that the parietal cortex enables people to be aware of their intent before movement, not whether that intent was first generated consciously. There is also the problem that the really crucial point was not tested; namely, can subjects distinguish between a stimulus-induced feeling of intent and an internally generated actual intent? There is also the point that brain activity is probably different between test conditions in which a subject randomly and spontaneously wills a movement and when such a movement is planned.

On the other hand, this work is a refreshing departure from Libet-type experiments. Because the focus is on stimulation, the limitations are of a different order. The authors did note the earlier research on the cognition of intention and the robotocist theory. But they were careful neither to endorse nor criticize the robotocist theory. Instead, they made the limited interpretation that realizing the will to move precedes movements and even intended movements that do not occur.

As will all such studies, I contend that the investigators only considered a subset of all the brain areas that are known to be involved in willed actions. For example, there were no electrical stimuli delivered to prefrontal cortex areas that are known to be involved in generation of intent and executive control in general. Just because realization of intent appears in the parietal cortex does not mean that intentions were generated there. Even so, whenever intent is generated, it clearly must precede the realization of intent, and the studies clearly showed that realization of intent can occur without movement.

Freewill studies have been plagued by unwarranted simplistic assumptions and circular reasoning. Robotocists commit at least twelve major fallacies of logic or acceptance of insufficient data in interpreting experiments of these kinds. I summarize these flaws as follows:

1. Increased neural activity has ambiguous meaning. Increased neural activity in a given brain area, such as the readiness potential in Libet's experiments, may not be limited to just one function. A relevant example is a study in which volunteers were asked decide, after hearing a tone, whether or not to tap on a keyboard.[45] The readiness potential was present regardless of their decision.

Another problem is that until recently nobody examined the electrical response of individual button-press trials, which are normally averaged over many repetitions. The readiness potential is not the same in each of the trials averaged in the fraction of a second prior to a willed button press. A recent repeat of the Libet experiment sorted the response from each individual trial.[46] Almost half the time the voltage was of opposite polarity, even though the button still got pressed. Most importantly, recording the potential when participants had to listen to a beep and refrain from pressing a button revealed similar potential shifts prior to the beeps, even though there was no willed button press. The difference was that in the beep task there were equal numbers of positive and negative potentials, and they cancelled out to zero on average.

So it seems clear that the readiness potential is an unreliable measure of decision making. Just what it reflects is unclear. Perhaps this potential reflects a nonspecific process of task expectation, facilitating an intended movement in anticipation of a decision made earlier elsewhere in the brain. That earlier decision could thus have been made consciously (that is, freely) long before the readiness potential or the conscious realization of the decision.

The readiness potential may just be a reflection of a neural competitive process in which activity builds up to reach an action threshold independently of what was driving the activity toward threshold (that is, causing the behavior). In fact, just such a process was observed in a repeat of the Libet experiment, wherein participants were asked to act as soon as they could only when they heard a click. The researchers predicted that a readiness potential would appear only in those subjects that responded quickly; that is, they would be the ones in which neuronal activity had reached the threshold even when the subject was not preparing to move at that particular moment. That is in fact what was observed, and, moreover, the slow-to-respond subjects did not show a readiness potential.[47]

2. Decisions are not instantaneous. What we consciously will could well be spread out over time. The process can be ongoing, but our realization captures the process only as a snapshot in time that suffices to label the decision but not the process. Moreover, in the typical experiments the subject thinks about the test situation continuously within the rules of the experimental conditions. The only thing at issue is *when* to act. Even the decision of when to act is not instantaneous. Even if not verbalized with silent self-talk, the subject has to monitor time and think consciously about what is an appropriate time to act. "Has too much time elapsed since the last act? Should I use a set pace of responding or use a semirandom pattern? How often do I change my decision to act now and defer it?"

In a more complex situation, conscious decision making is even more obviously an ongoing process. We weigh the evidence. We lean one way, then the other. Finally, the preponderance of evidence and the weights we assign to it lead to a conscious decision. The decision itself may have been instantaneous, but its process was dominated by choices spread out over days, months, or even years.

3. Conscious realization of intent is not instantaneous. Just as it takes some time to make a decision, it takes some time to realize you have made it. Libet cited one of his earlier papers that showed that conscious realization is not a point process; it takes at least a half second to be realized. In human subjects who were electrically stimulated in the body-mapped sensory cortex, the stimulus had to be delivered for a half second or longer before they realized the sensation.

In typical freewill experiments, two conscious realizations have to be accomplished at more or less the same time, neither of which can be assumed to be instantaneous. In addition to realizing when a willed decision to move has occurred, the subject also has to realize consciously and hold in working memory the position of the clock indicator. To do this, does the subject think about the clock in the context of "I am about to move and must make sure I note the time" or does the subject force a spontaneous movement and then switch attention, after significant delay, to realize the time? Both the decision and the time recognition need external validation. A definite amount of time is clearly required for both realizations.

When images are presented in sequence, for example, it takes around a tenth of a second to accomplish the correct conscious recognition of an event.[48] In other words, subjects need this time after seeing an object to consciously process what it is and what category of objects the image belongs to.

Compared to the time needed to categorize an object, on average, an additional sixty-five thousandths of a second were necessary for identification of what the object was. Using visual images to test the time for conscious recognition of an event generates especially useful evidence because vision is an exceptionally high-speed process in the brain, very likely to be much faster than the conscious processes needed in a Libet experiment, where one must decide to move, determine what to do, and decide with what body part to do it while being consciously aware that these events have occurred. In other words, you can make a conscious decision to act, but it may take you several hundred thousandths of a second to realize what you have done.

4. Decision making and decision realization are separate processes. This could impose delays because decision making is accomplished via numerous synapses in widely distributed circuits whereas the movement command can be executed via as few as two or three synapses. It is also possible that conscious realization processes are not complete until they are confirmed by feedback from seeing and feeling that the movement has actually occurred. Realization captures the process as a snapshot in time, but the antecedent spread-out process of realization goes unrecorded.

5. Decision making is not the only process going on. Actually, there could be five conscious processes going on prior to movement commands in the Libet-designed experiment. In such experiments, the brain must say to itself the equivalent of

> "I know the rules of this game and agree to play by them."
> "I intend to move soon (and withhold movement in the meanwhile)."
> "I realize and confirm that I have issued the order to move."
> "I notice the time I issued the order to move."
> "I hold this time in my working memory so I can report it afterward."

Decision making is a process, not a discrete event. The same principle applies to unconscious decisions, but I make the point here because robotocists seem to overlook the role of multistep processes in conscious decision making.

Let us recapitulate what must be happening during a conscious decision to make a movement in a real-world context. External stimuli or even internally generated signals would generate a conscious decision to perform a given act. These signals must activate memory banks as a check on the appropriateness of the movement in the context of what has been learned about making such a movement. The reward system has to be activated to assign value to the making of such a movement, weighing the expected immediate utility with the longer-term value. The emotional networks of the limbic system have to be activated to see what level of passion, if any, is appropriate to the movement. Movement control networks have to be activated in order to plot a trajectory and to evaluate the correctness of the anticipated movement. There are "premotor" areas of cortex that are engaged in planning for the movements that are to be executed.

6. Not all intentions are for simple movements. There is also the issue of the kinds of movement we wish to associate with conscious intent. In speech movements, for example, we have all experienced high-speed conversation, clearly controlled by conscious intent to express thoughts, both spontaneous and in response to what is said by others. Consider all the thoughts one has to hold in conscious working memory to conduct intelligent conversation. We think consciously about what is in working memory as we use it.

Studies of the Libet kind bear no obvious relevance to real-world situations in which people use and are aware of prolonged deliberative processes. The deliberation clearly occurs in the consciousness, and it most likely occurs before a final decision is made. Are we to conclude that the results of conscious deliberation are somehow sent back to unconscious mind to let it make the decision?

7. Not all willed intentions are formed in acts of decision. Especially in the case of habits, decisions are made long before the execution of an act. Mele points out that an intention to do something can arise without being actively formed from a current decision process.[49] Not only are some habits originally formed consciously, but the choice to deploy a habit may be made consciously and certainly, as Libet suggested, be vetoed consciously.

8. Conscious decisions can be temporally uncoupled from the action. I may decide this morning, for example, to be more thoughtful toward my spouse. Opportunity to do that may not arise for hours, as for example, when I come home from work that evening. When the opportunity arises that evening to be thoughtful, do I have to remake the decision? No, it had already been made hours ago. One could argue (but not test) that the evening's behavior was generated unconsciously, but it could not have been driven by the process of making a current decision because that had already been done.

9. Lack of high-resolution time monitoring of the decision process (when it starts and when it ends). While freewill experiments based on electrical recording provided high-resolution time monitoring, there is no corresponding high-resolution indicator of conscious decision making. Surely, there are electrical indicators of conscious decision making, probably scattered throughout the brain, but unlike the readiness potential of the motor cortex, there is no time marker for when the intention and decision processes start and end.

Nobody knows where in the brain the conscious self is, probably for the same reason that we can't find where many memories are stored. The phenomena are not located in a place; they are processes in a population. That is, they are processes that occur in the form of patterns of neural activity distributed across a large population of neurons.

Also, the motor cortex only *begins* its increased activity before the self-reported intent to move. Few analysts admit how little we really know about what is signaled by this readiness potential. In the early ramp-up of the electrical signal, the change could indicate a preparation for later movement, or that a movement command was about to be issued, or that there was intention to move in some nonspecific way. That intention likely is generated elsewhere. Maybe the processing of intention triggers the ramp-up, making it inseparable from the general instructions in the experiment.

Moreover, some aspects of the decision are predetermined. Subjects have already decided to generate an urge to move the finger sometime during the rotation of the clock hand, long before the start signal of the "urge" to move. Of necessity, the subjects were holding these instructions in conscious working memory. So in that sense subjects had already consciously planned and willed to move long in advance of selecting between the options of move or don't move. The only part of the decision remaining was when to do it.

10. Inappropriate reliance on self-reported awareness of actions and time-estimation accuracy. In self-reported awareness of a conscious decision, the issue is determining whether (a) the intention occurred prior to the realization of having made a choice or (b) the awareness was reconstructed after the action occurred. I only need to mention a few studies to make the case that humans are not precise in their awareness of time compared with actual time on a fraction-of-a-second scale. One study, for example, showed that subjects made major errors in time estimation when instructed to keep visual displays on a screen for a fixed time.[50] Moreover, the accuracy was affected by prior priming experience with the images.

A review of a variety of reports shows that time-estimation accuracy is affected by experimental conditions, such as stimulus modality, degree of attentiveness to time, and level of arousal. Authors of the review showed in their own experiments that time estimates were affected by prior expectations about visual stimuli.[51]

Stanley Klein replotted Libet's original data and found that observers had great uncertainty about the relative timing of events. He also pointed out that the Libet design required responses whose timing was difficult to judge.[52]

Several experiments document that it takes time to process visual information consciously. In an experiment originated by Romi Nijhawan, subjects assessed the timing of an object passing a flashbulb. The timing was exact: the bulb flashed precisely as the object passed. But subjects perceived that the object had moved past the bulb before it flashed.[53] This suggests that the brain projects a moving event a split second into the future, seemingly working on old information. Apparently, the brain needs time to consciously register what the eye sees. In the context of a Libet-type experiment, realizing the location of a clock hand occurs after the time of the actual decision.

As another example of timing error, when subjects were asked to make subjective timing decisions about a stimulus, they consistently tended to report events as happening about seventy thousandths of a second later than the events had actually occurred.[54]

Another recent study of time-awareness accuracy used the control condition of the Libet method and required subjects to judge the time of occurrence of a stimulus relative to a clock indication of time.[55] Response accuracy varied systematically with the sensory modality of the stimulus and with the speed of the clock. If these indicators of externally observable events are inaccurate, the researchers suggest that the time estimation may also be inaccurate for endogenous events.

Awareness of time is only one indicator of how well humans are aware of their actions, and it can be argued that humans have awareness limitations that go beyond time awareness. For example, a recently published paper reports that awareness of our actions depends on a combination of factors involving what we intend to do and what we actually did. One interesting experiment required subjects to reach consciously for a target that jumped unpredictably on some trials.[56] Subjects were to express their expectation of a target shift, point at the target as fast as possible, and reproduce the spatial path of the movement they had just made. The last step of reproducing the trajectory was taken as an index of the awareness of the previous action.

The accuracy of reproducing the trajectory was measured in terms of the degree of movement undershoot or overshoot. On trials where subjects thought there would be a target shift, the overshoot was greater and the undershoot was less than on trials with lower expectancy. Thus, conscious expectancy affected the awareness of what had taken place.

Time-awareness accuracy is confounded by the likelihood that the whole process of decision making and monitoring has many elements that combine unconscious and conscious processes. The time scale used in Libet-like studies is too short to adequately capture all conscious processes. In the Libet study, the actual movement did not occur until after subjects thought they had decided to move, which allows for the possibility that the processes above could have participated in a conscious will to move. Some portion of these processes occur at an unconscious level that could have primed the motor cortex to start a readiness ramp up of activity to await final confirmation from conscious decision making.

And how do we explain other kinds of decisions that are so rapid that long preparation periods are not possible? For example, in high-speed conversation, the brain has to process (a) what a listener hears in the context of what he or she wanted to say, (b) how those thoughts have to be modified to make an appropriate response, all while (c) anticipating how the other person will react. Because of the very high speed of this processing, it is unlikely to have originated unconsciously, followed by the otherwise needed delay in priming the motor cortex to make the appropriate tongue and lip movements for rapid speech. To do this at high speed, however, requires that conscious processes have limited moment-to-moment information carrying capacity, which is the basic reason that working memory has much lower carrying capacity than unconscious processes.

11. Unwarranted extrapolation to all mental life. Even if unconscious choices are made prior to conscious awareness in a certain task that is not proof that all mental life is governed this way. How can this kind of methodology possibly be appropriate to test whether consciousness caused a given decision on an optimal plan, a correct problem solution, an appropriate conclusion, the appropriate interaction with others, which words to use in conversation, or what attitudes and emotions to embrace?

Complex tasks are probably performed in different ways than simple ones.

It may be that the reflex-like button press response is so simple that it can be caused unconsciously, as there is little need to assign or recruit assistance from conscious mind in making the decision.

12. Conflicting data or interpretations are ignored. Recall the data of Soon's group, which showed increased activity in two regions of the frontal and parietal cortex a full seven to ten seconds before conscious awareness. This was considered evidence of unconscious motor preparation. There is no basis for believing it takes ten seconds for unconscious mind to prepare motor pathways for a button-press movement; the brain issues commands much faster than that.

Why do robotocists assume this predictive change reflects motor preparation instead of the processing of conscious choice and other cognitive functions associated with the "rules of the game?" These areas of brain normally have conscious functions and not movement functions.

Consider rational thinking in the context of the freewill issue. Of course, I suppose we could analyze a situation, compare and contrast with other situations, determine salience and value, and arrive at conclusions without the aid of conscious thinking. What seems to be conscious editing and executive top-down direction of a rational process could all be performed unconsciously, with consciousness being periodically informed of the unconscious editing and direction. But doesn't that seem like a convoluted way to explain rational thinking when the alternative of conscious executive control is a more direct and obvious explanation?

Why doesn't irrational thought get corrected unconsciously if unconscious processes correct rational thinking? Well, I can't prove any of this one way or the other, but certainly simple-minded button-press experiments do not support the notion that rational thinking is an example of illusory free will.

To conclude from simple button-press experiments that conscious mind only observes, not intends, decides, or acts, requires a leap over a tall mountain of objections. To me, what it shows is that even the supposed smartest people on the planet can jump to conclusions just like the rest of us. It is now increasingly clear that scientists and philosophers have built their illusory-freewill edifice on a foundation of quicksand.

* * * *

Robotocist bias may even keep investigators from looking for evidence crucial to the argument; namely, *neural representation of intention*. Yet there is enough evidence to indicate that there are neural representations of intention, as for example in the Desmurget study.[57] A slow time scale allows for conscious awareness of intent, development of plans, and "on-the-fly" adjustments. Consciousness allows us to reference what we know and think about the future so as to anticipate explicitly what we need to do to get what we want and to plan accordingly. Such intentional planning has a neural representation and can even be detected experimentally in animals. In one such study, investigators monitored neurons in a planning area of the monkey brain.[58] They put electrodes in an area of neocortex that was known to be required for planning, but not actually making, arm movements to reach a target. The planning area in monkeys is a small patch of neocortex just above the ears. Monkeys were trained to "think about" a cue presented on a computer screen that told them to plan a movement toward an icon that had just flashed on a screen in one of up to eight locations. Each location was associated with a certain firing pattern in the planning neurons. Here is a clear case where the intent to do something was established long before any action occurred because the monkey was planning how to carry out the intent. While monkeys thought about the required movement, computer analysis of the firing patterns of these neurons could predict what the monkey was intending to do—tantamount to reading the monkey's mind ahead of the action. The researchers knew that it was intention that was represented, not actual movement or even planning for movement, because the monkeys were trained to get a reward only when they withheld actual movement but nonetheless made the correct planning, as indicated by their neural firing patterns. It is clear that these conscious monkeys had a mind that contains neural representations for decision processes, and that these representations are active prior to planning how to control movement.

If consciousness causes action, there should be some neural correlates when such will is being exercised. Almost certainly, consciously driven decisions emerge from distributed processes in the neocortex. One might monitor multiple neuronal activities within appropriate areas of the cortex. For example, if the willed task involves vision, multiple areas of the visual cortex should be monitored. Perhaps changes in impulse onset/offset, firing rate, or sequential-interval patterns would be seen in certain neurons. Perhaps there would be changes in oscillatory frequencies of field potentials or in coherences with oscillations elsewhere or with other frequencies.

Degree of synchronization can be frequency specific, involving shifts among various brain areas and even oscillators of different frequency. As my ambiguous figure studies revealed, when subjects made a conscious decision about which image was perceived in an ambiguous figure, there was significantly increased synchronization across widely distributed scalp locations.[59] We also found, much to our surprise, that synchronization occurred in multiple frequency bands, a finding that has also been reported by others.[60] The obvious interpretation is that these synchronization changes are correlates of conscious choice in what to perceive. They are also correlates of willed choices; that is, subjects could decide what percept to hold in consciousness, as for example a vase or a face.

In fact, for many such images, many subjects have to extend considerable will power to perceive an alternative image because their default percept is so strong. Since oscillatory synchronization is so tightly associated with this process, this may be the clue that conscious choice is enabled by synchronization of certain oscillations. An experiment could readily check for changes in coherence patterns when one chooses to hold the difficult percept in consciousness as compared with patterns during the default percept. The experiment might benefit from including a time indicator, of the Libet or Soon type, for when subjects realized they wanted to force perception of the difficult alternative image. If synchronization changes indicative of intent occur after the subjects' indication of conscious intent, it might support the robotocist hypothesis. However, we would still face many of the faulty assumptions mentioned earlier (intent processes are smeared in time, extra time is needed for realization of intent vs. generation of intent, and so forth). The emphasis in analysis should be on correlating electrical activity with mental state, not on precise timing of which came first.

Another step in the right experimental direction is the experiment reported in 2004 by Daeyeol Lee at the University of Rochester.[61] He monitored the level of time-locked oscillations in electrical activity in the cortex of monkeys during a task in which they made a predictable series of hand movements as they integrated sensory signals with expected reward. Movement performance was influenced by both the position of movement and the location of the rewarded target, but only the consciously expected reward affected the degree of synchronization. Synchrony of neuronal activity clearly seems to be a marker of something different from the amount of activity.

In one promising new kind of study, a group in the Netherlands examined unconscious inhibitory control and found that the prefrontal cortex controlled the inhibition.[62] This region of the cortex is well known to be critically involved in conscious decision making. The group used a Go/No-Go task, while at the same time using EEG evoked-response monitoring to track the brain fate of the cue that served as a No-Go inhibitory signal. Other researchers had reported that such cues produce an evoked EEG response and that the EEG waveforms clearly distinguish between Go and No-Go trials.

Human subjects were to respond as fast as possible to a visual Go signal but to withhold their response when they perceived a No-Go cue preceding the Go signal. The No-Go signal was a visual cue that was presented faintly enough that the cue was not consciously perceived when the cue was presented near the time of the Go signal but was perceived consciously at longer intervals. With and without conscious perception of the No-Go cue, the task could be performed correctly. This does support the notion that unconscious processes can drive conscious behavior.

The EEG revealed that No-Go cuing was associated with reproducible EEG responses over the prefrontal cortex. However, what is important is that the EEG was different depending on whether the No-Go cue was consciously perceived. Thus, consciousness may have its own decision-making process. This still does not rule out the possibility that the EEG sign of conscious involvement is simply reflecting the process of conscious realization of an unconscious decision. Maybe neuroscience will prove unsuitable for testing whether free will is illusory.

What about the core issue of how a conscious mind could freely generate choices and action? If thinking is the flow of CIPs, how does that enable us to make intentions, choices, or decisions? As a physical representation of the current mental state, CIPs can be changed by external input or by feedback as the CIP codes are routed and modified through various circuits. And let us remember what intentions, choices, and decisions really are. They, too, are CIP representations that can be propagated to generate new thought, intent, choice, or decision.

When decisions are performed by the conscious mind, the avatar can differentially select what inputs from stimuli and memory to consider, thus ultimately biasing one option over others. If conscious mind is an avatar, acting as an agent with a "mind of its own," what is to keep it from making its own choices? This might be the basis for free will.

A main reason for all the dispute about illusory free will is the wide range of definitions of free will. At the heart of the robotocist stance is the physics perspective of quantum determinism. The doctrine of determinism states that every fact in the universe is guided entirely by physical law. Quantum mechanics also posits that nature is random until observed or measured. Neither facet provides much room for free will in human decisions. Fatalism and destiny governed by probabilities are corollaries.

However, the equations of quantum mechanics do not predict what will happen, but rather the probability of what will happen. Human behavior is much more complex and subject to numerous variables that have nothing to do with quantum mechanics.

Free will is greatly removed from the activity of subatomic particles. A more practical criterion for free will is that of "could have done otherwise." Choosing one among a set of alternative choices is considered free if nothing forced the choice and you could have made an alternate choice. Thus defined, quantum mechanics seems irrelevant and robotocism is hard to defend.

Here's a useful analogy about free will: we are all strapped to a ball and chain just by our human condition. However, our freedom exists inside the circumference within which the chain allows us to move. In other words, free choice can exist even when there are constraints. Humans operate within a finite range of probabilities, formally called a *probability density function*, for any particular choice or act. Each potential choice or action has its own range of probabilities. The odds, for example, that you will have eggs for breakfast this morning may range from 0 to 40 percent, depending on such constraints as whether you have any eggs on hand, how much time you have to cook them, what you ate during your immediately preceding breakfasts, and so on. Your choice is open yet constrained within those limits. Free will does not have to be completely free.

Folklore, common sense, and psychology all implicate the conscious mind in mediating, programming, and even controlling much of what the unconscious mind does. All of us avatars make choices that are not forced on us. We think that many, if not all, of these choices are made freely. As a child, I chose to play with kids I liked and avoided those I did not. Nobody forced me to pick one over the other. Even when we chose sides for a pick-up ball game, I chose the kids who I thought were the best players among those not yet selected for the other side. I chose my wife. It was not a shotgun marriage. At different times I chose different career activities and rejected the pursuit of others. Robotocists would claim that each choice was inevitable. How can they prove that?

I once read a specious argument about coin tosses, where the robotocist writer supposed an alien observed that sometimes a coin comes up heads and sometimes tails. The alien concludes that the coin is alive and can "decide" which way to fall. We, of course, conclude that the coin toss is governed by the laws of physics and probability because we know the coin is not alive. But if the coin really were alive, it might have a mind of its own and might make choices that seem to violate the laws of probability. The fact this does not occur just proves the coin is not alive.

The writer claimed that our brains are like that coin toss, made of organic matter rather than metal. Therefore brains must, like the coin, lack free will. That is a false analogy. Moreover, humans certainly do improbable things that are not predictable.[63]

Usually, when we are consistent in making certain choices in the face of other equally likely alternatives, it is because of something our brain has learned. Sometimes it has been learned by the avatar, in which case the repeated choice may become an automated habit. That should not obscure the fact that the early choice making was freely chosen, not unconsciously automated. In such cases, the avatar has learned that the favored choice has more value or is more useful than the alternatives and so the avatar has no further need to make an alternative free choice. Just because a given choice is favored doesn't mean it is not freely made.

Some critics like to say that choices are made by neurons, and that neurons operate by the laws of physics. What is the law of physics that made me like one kid and avoid another, or made me choose the particular wife I chose to live with for forty-nine years before she died? What makes me write this book?

Neurons can be doing lots of different things at any given moment. They even sometimes do things spontaneously, what you might call random. But as Cliff Sherry and I and many others have shown, impulse trains often contain serial ordering of their intervals, not a random distribution. When neurons do something purposeful, as in directing a willed action, they do it in the context of what the neurons have learned in the past and in the context of the momentary situational contingencies. That decision may take one of perhaps several alternative directions. The direction chosen may be random or inevitable (thus not free) or may be repeated consistently because of learned appropriateness. Surely, we shouldn't say that learning strips one of any chance for free will. In fact, at each learning experience, one still has the option of accepting or rejecting the information in that experience. One can also choose different learning experiences.

Examples of Apparently Conscious-Driven Action

Numerous commonsense examples could be constructed to illustrate complex situations in which conscious intent is likely driving behavior. The examples I give are all based on presumed conscious will to make certain movements that are much more demanding than a button press.

Here is one example: you are driving a car in heavy traffic and another car runs a red light, pulling into your path. You can realize the full nature of the emergency and intend to turn the steering wheel appropriately and move your foot off the accelerator and onto the brake pedal long before you actually make such movements. You may not be able to avoid the accident that you consciously intended to avoid. The conscious analysis of the emergency, the intent to make certain movements, and the motor execution are all completed in a fraction of a second. And we need to take into account the fact that a conscious decision can be made but not realized for up to half a second. It doesn't seem likely to me that all this was figured out unconsciously, then conscious awareness was engaged, and then conscious awareness was realized in that same fraction of a second (remember it takes about half a second to realize you have made a conscious decision). How can the responses be generated unconsciously when the unconscious has not been fully preprogrammed for such movements? From beginning to end of the episode, conscious-intent processes are clearly operative. Without conscious adjustment you would surely have a wreck.

Here is another example, this one relevant for football fans: in almost every game there is at least one play where a pass receiver drops the ball because he was consciously thinking not only about catching the ball but also about the moves he would make after the catch because he heard a defensive back thundering toward him. All this was going on in conscious mind long before the brain issued the movement commands needed to catch the ball and cushion the impending blow. You might argue that the preparation to move was triggered before all the conscious realizations about the pass-receiving context, but that can't be measured. As in the car-accident case above, there is no way the unconscious is preprogrammed to make all the right movements, given all the variables involved and the uniqueness of every pass-catching challenge. In any case, it seems clear that conscious thought and decisions were being made well before complex motor commands were issued and modified in the last few milliseconds to adjust to the ball's trajectory and speed so as to accomplish the desired movements.

True, intent to move probably is preceded by unconscious enhanced muscle tone and preparations and rudimentary alternative sets of muscle commands that could be considered for movement. But it is hard to argue that conscious thought about how and when to move is preceded solely by unconscious processes. Conscious planning, by commonsense definition at least, always precedes delayed action. Robotocist scientists will point out that commonsense can be wrong. But so can scientific dogma.

If unconscious mind does everything, and conscious mind is merely a reporter that may intervene on occasion, we have a problem in explaining the decisions and conclusions we make regarding

- attitudes and beliefs we choose as a result of introspection;
- conclusions we reach based on literature, poetry, art, or music;
- decisions about which words to use in rapid conversation;
- choices we make about time (past, present, and future);
- intentions we use in early-stage learning, such as when riding a bicycle or touch typing;

- decisions about what to believe when considering politics, religion, and so on;
- decisions about taking or avoiding responsibility;
- decisions about whether to lie and deceive;
- choices that emanate from conscious analysis;
- choices made in developing plans for the future;
- feedback adjustments to ideas, attitudes, emotions, and behavior.

The unconscious mind surely participates in all these human-thinking activities, but to presume that all of these activities are governed only by unconscious mind is an assault on human reason. To make such an argument, you have to assume that consciousness can't do anything. But consciousness does many things, in my view (see below).

Only a few scientific studies of these issues have been performed, and each has involved only decisions to make simple movements that one already knows how to do. As I mentioned, these studies suffer from seriously flawed assumptions and interpretations. Also, each of these studies is contaminated by the requirement of prerequisite processing needed to hold in conscious working memory the situational contingencies and any ruling constraints.

In other words, robotocists have jumped to conclusions—hardly a judicious scientific stance. Until science provides evidence (as opposed to speculation cloaked in pseudoscientific garb), it is scientifically irresponsible and dogmatic to insist that there is no such thing as free will. It seems to me that such scientists are left with weak arguments from authority, as indicated by their citing Darwin and Einstein as robotocist allies.[64]

I have not seen anyone make the following point: in learning a new skill, such as playing the piano, there is no way the unconscious mind can control movements in the beginning because it has no way of knowing what to do. It does not know, for example, which keys are in the C chord until conscious mind informs it. Only the conscious mind can choose which keys to press because initially only it knows what should be done. If that is not free will, what is? Apply this principle to many of the things you learned as a child, like tying shoelaces or riding a bicycle.

Moreover, I have not seen freewill deniers take on the issue of creativity. By what evidence can they conclude that creative thinking is preprogrammed? We surely must concede that much of brain operation is stochastic. That is not to say it is random or without cause, but rather that the cause may not always be predetermined.

Now we come to the tipping point of how the brain selects among stochastic options, each of which may have its own probability-density function. Surely, not all choices are made randomly. And surely, they are biased by prior programming. But it would seem that the brain is capable of making a new (creative, if you will) choices that are not compelled by programming. We know from common observation that some brains are highly creative. The corollary is that some people have more free will than others. I know lots of people who are examples of that.

Children are especially effective at deploying creative imagination. Fantasies and new ways of thinking come easily to them. As a result, scientists have been using fMRI brain scans to learn more about how children do this. In one study, scans were taken of fifteen children who were asked to imagine abstract visual shapes and mentally combine them or dismantle them into separate parts.[65] During imagining, widespread activation occurred over a large part of the brain, indicating that creativity engages a large part of the brain's "global workspace."

This research needs to be interpreted in the context provided by the fact that children are more creative than adults and, of course, in the context of free-will debates. The findings suggest that children are especially adept at accessing more of their global workspace. Why can't adults do that as well? If creativity is programmed by past learning, adults should be more creative than children, as children have little knowledge upon which to build mental programs. This is consistent with the common observation that scientists often become more creative when they switch fields or topics within their field. It is as if accumulated information and experience get in the way of creativity and reduce the brain's ability to access enough network workspace to be fully creative. Certainly willed creativity is more freely done in children and suggests that the brain circuitry of children is less walled off from programmed access than in adults. But children don't have much programming. In other words, I would suggest that prior programming does not enable creativity but rather diminishes free choice and the capacity to achieve it.

If you insist that free will does not exist, then don't you have to insist that no brain is creative; that is, that no brain can generate thought, decisions, choices, and plans that are not compelled, programmed, or emergent from some probability-density function? In short, I see the need to prove that creativity is preprogrammed if you want to explain why free will is illusory? Is creativity also an illusion?

The perspective I want to emphasize is that everyday intentions, choices, and decisions can arise through a combination of unconscious and conscious actions. Which mode of operations prevails, robotocist or freely driven, depends on the nature of the situation and the individualized brain. The idea is that for simple, well-learned, or habitual tasks, the unconscious mind issues the intent, choice, or decision and the conscious mind then realizes what has been done, which in turn may or may not lead to doing anything about it. For complex or novel tasks, however, the conscious mind does the processing and informs (programs) the unconscious mind, which likewise may or may not do anything about it. The brain (mind) is an autonomous decision maker. The criteria by which its decisions are made may well be programmed, but that same mind can choose the experiences and learning that does the programming. Why would anybody be surprised that such an autonomous essence could have free will? Surely, if brain can choose its own programming, it can—on the fly—even choose to modify its decision criteria. I would call that free will.

In the real world, unconscious and conscious minds interact and share duties. The unconscious mind governs simple or well-learned tasks, like habits or ingrained prejudices, while the conscious mind deals with tasks that are complex or novel, like first learning to ride a bike or play sheet music.

We do act like zombies driven by our unconscious when we act out of habit, prejudice, or prior conditioning. But we should and can be responsible for what we make of our brains and for the choices in life we make. In a freewill world, people can choose to extricate themselves from many kinds of misfortune—not to mention make the choices that can prevent misfortune.

Doubts about the existence of free will should fade when one thinks of the conscious mind as an avatar. Because the avatar is materialistic, it has the same means as unconscious mind to influence and be influenced by what the rest of the brain's circuits are doing. In other words, a conscious avatar being can intend and choose because it operates with CIP mechanisms. It surely has more degrees of freedom than unconscious mind.

The Purpose-Filled Life of Consciously Driven Action

Think about the implications that follow from the robotocist ideology that claims consciousness can't cause anything. If consciousness can't drive thought and behavior, what good is it? Do we really think that conscious purpose is not consciously generated? What good does it do to be aware of our pains and pleasures if there is nothing that such consciousness can do about them? Maybe the robotocists will at least concede that consciousness can inform the unconscious mind so that the unconscious mind can use the information in its generation of purpose and willed behavior.

There is abundant evidence, anecdotal and experimental, that suggests that the unconscious mind makes us do things that we really know we should not do (as in, "the devil made me do it"). But it seems patently absurd to extend such observations to the conclusion that we have no free will or that free will is inevitably overwhelmed by unconscious demons.

How can something as "impersonal" and physiological as CIPs have any kind of purposeful will, much less *free* will? Let us recall that "will" is little more than an intent that is often coupled with bodily actions to achieve the intent. Intents exist as specific CIPs. The neural circuits automatically generate actions in response to conditions that call for a response. Such actions are stereotyped and inflexible whenever they are controlled without conscious oversight.

Willed action, whether robotocist or not, usually has a purpose: we purposely choose to do or not do specific things. We avatars exist to extend the purposes of our unconscious minds. To the extent that free will exists, the avatar is provided the only form of freedom that is truly available. All else is constrained by physical, biological, and social contingencies.

Assuming that avatars have some autonomy of their own, they can develop their own purposes. Unconscious purpose is limited because it is not freely chosen. Such purpose is driven by hard wiring (as in knee-jerk responses), past learning, emotions, and other biological exigencies.

Avatars, however, because they are consciously self-aware, can generate their own purposes. Biologically speaking, the capacity for conscious action means that humans can rise above unconsciously driven behaviors that are unwise or maladaptive. The greatest value of the capacity for conscious choice is that it gives people power over themselves and their environment. Just believing in conscious choice has consequences. Maybe the belief itself is an act of conscious choice. What overwhelming evidence requires robotocists to argue that specific beliefs are inevitable?

Avatars can also recognize that their purposes can be biased by unconscious influences. Avatars can introspectively examine the instincts, drives, and prejudices that otherwise would determine choice and action. Avatars can veto bad decisions of their unconscious minds.

Just having the capacity for making choices freely can have a purpose of its own. Several formal studies show that belief in conscious choice matters. One study showed that the level of such belief affected performance of students taking a test for monetary reward.[66] Students inclined to disbelief in free will were more likely to cheat on the test. Another study, in which level of belief in free will was manipulated, revealed that inducing people to be more doubtful about free will make them more aggressive and decreased their helpful, prosocial behaviors.[67]

More recently, a study showed that job performance is affected by a belief in free will.[68] In a first experiment using undergraduates (mostly women), free-will belief affected how students rated their expected success in future jobs. Correlation of scores on a freewill belief survey and ratings of expected career performance showed a strong correlation. Other factors were also correlated, such as the student scores on tests for conscientiousness and agreeableness. The other factors tested, extraversion, emotional stability, and SAT scores, were unrelated to estimates of success. However, belief in free will was independent of the other factors, predicting expected career success above and beyond the other two positive factors.

A second experiment evaluated people already in the workforce and compared freewill beliefs with actual job performance, as measured by actual supervisor ratings. Level of belief in free will strongly correlated with overall job performance. No other tested independent variables were relevant: these included life satisfaction, work ethic, and personal energy.

Of course, correlations do not necessarily indicate that one thing causes the other. But common sense suggests that if you believe you can change your life for the better, you should be more likely to do the things necessary to create success. Disbelief in free would logically have the opposite effect.

The Role of Learning and Memory in Conscious Choice

Neurons can be doing lots of different things at any given moment. They even sometimes do things spontaneously, what you might call random activity. But when they do something purposeful, as in directing a willed action, they do it in the context both of the momentary situational contingencies and of what the neurons have "learned" in the past. That decision may take one of perhaps several alternative directions. The direction chosen may be random or inevitable (thus not free) or may be repeated consistently because of learned appropriateness. Surely, you don't want to say that learning strips one of any chance for free will. In fact, at each learning experience, one has the option of accepting or rejecting the information in that experience.

One of the things neurons may learn is to habitually accept defeat. That is, they have learned to be helpless.[69] Learned helplessness has been well documented in both animal and human studies. Very young elephants, for example, can be trained to stay tethered by staking their restraint chain so that escape is impossible. Years later you can stake them out with a simple wooden stake that they could easily pull out, but they don't even try. In people the situations in which learned helplessness develops are much more complex, but the effects are similar. Usually, these are situations in which a person concludes that there is little that he or she can do. Trying harder or using a different approach is perceived in advance to be doomed to failure. The doubt breeds powerlessness. Even though the doubt may be irrational, it still has the power to immobilize us. Commonly, we feel not only that we are unable to escape our fate, but also that we are victims.

The learned element of such helplessness is key to its creation and to its cure. Past lack of success creates the state that chains us. The cure is conscious reasoning that unmasks the irrationality of learned helplessness. With force of will, we can take action and choose to pursue alternative goals, strategies, and tactics. We do not have to be chained to the stakes of the past.

If people believe that they have no control over their lives, why would they even try to make things happen? People who don't believe they are in control make excuses and attribute their situation to fate, bad luck, or powerful forces and people beyond their control. My *Blame Game* book was designed to show such people that they have learned how to be helpless, and that they can likewise learn to help themselves.

In *The Power Principle*,[70] Blaine Lee begins with the premise that human learned helplessness is a matter of choice, adding that this is at the root of lack of personal power. When you choose to be powerless, you ignore, disregard, procrastinate, neglect, dismiss. The consequences lead to living with the status quo of anxiety, fantasies, diminished capacity, and helplessness. Lee's remedies begin with conscious assessment of just how helpless and powerless you really are. The state may be real, but typically it is because you have boxed yourself in and have not considered looking for a way out. Like the elephant chained to a wooden peg, you have come to accept your state as the norm, when in fact this is not justified.

There are alternatives, and these revolve around freely willed development of personal efficacy. This begins with a sense of agency, namely, that we are the agents of change. In Albert Bandura's book *Self-Efficacy: The Exercise of Control*,[71] cultural evolution is seen as moving from a collective sense of dependence on the gods (early humans relied on conciliation rituals directed toward the gods to improve their lot in life) to a realization that people have the capacity to shape their own destiny through free will.

Personal efficacy is seen by psychologists as a state of control over one's life. People clearly differ in efficacy capability, but people do have choices, and many are very effective at choosing wisely and following through to accomplish their goals. It is hard to see how all this can be accomplished without a substantial degree of conscious action.

What usually gets left out of discussions about conscious agency are questions about how a brain establishes stored-memory preferences and the role of conscious evaluation of extant contingencies in the context of such memory. These functions are surely causal, but the cause of the cause comes first. Any given brain is free within certain limits to consciously choose its learning experiences and what it will store as lasting memory. Those choices in turn are often governed by what a brain has learned about the self-interest value associated with given contingencies.

So, in some sense, it is learned values that underlie much of conscious action. It is the brain that assigns value. The real question is whether values are freely chosen or imposed. Values and beliefs are to a large extent optional and formed consciously. The conscious brain directs the choices that govern value formation and reinforcement.

Now, let us consider the positive situation in which people learn to make conscious decisions based on previous learning from other situations. This learning occurs in the context of the learned sense of self. The conscious brain is aware that it is aware of choice processing, and it makes decisions in light of such understanding. When a given alternative choice is not forced, the conscious mind is aware that it is not obliged to accept any one choice but can select any one of the available options. Such realization might even guide many decisions at the unconscious level. In either case, the probable value of each alternative is weighed in neural networks, which collectively reach a "decision" by inhibiting networks that lead to less favored alternatives. Thus, network activity underlying the preferred choice prevails and leads to a selective willed action.

Consciousness could be a special adaptation that makes us the superior learners that we are. Consciousness guides development of intent, focus of attention, plans, and belief systems, as well as suppression of unwanted distractions and integration of information across past, present, and future epochs. In short, consciousness amplifies the programming of the brain from sources both internal and external to the brain.

Consciousness makes our thinking explicit. The conscious mind not only modulates decision making; it also generates intentions. By making analysis and decisions consciously explicit, we can more effectively write the programs of the brain that change who we are, what we think, and what we do—now and in the future. To some extent, our brains can even rewrite the record of its past.

Consciousness enables introspection. How does reflection help us to make choices? Relevant past learning experiences are called up into conscious working memory where they can create preferences for certain choices that past experiences have shown to be useful. The ongoing reflection processes are mediated by network activity patterns that can alter the option representations. Reflection processes going on in neural networks are modulated by existing contingencies and the likelihood of benefit or reward. Values are not only learned, they are subject to change by new learning.

Brain circuitry assigns value, and conscious choice of learned values is largely optional. The conscious brain directs the choices that govern value formation, reinforcement, and preservation in memory.

Now we are confronted with explaining how neural circuit impulse patterns representing the sense of self can make conscious choices based on learned values. First, each person's conscious avatar is an active agent that not only makes decisions but also learns and remembers things that help it to make future decisions. What is learned is stored in memory as facilitated circuits and deployed "on-line" in the form of CIP representations of what was originally learned. The avatar is released in wakefulness from its stored representation and is able to make its own decisions using what has been previously learned. Avatar processing is certainly not random, and it presumably can occur with a greater degrees of freedom than is found in the unconscious mind.

To move from helplessness to personal power, one must consciously learn that the discomfort and inconvenience of being helpless is greater than the difficulty of making a change. Change requires a kind of cost-benefit analysis to guide rational choice. Once a decision to change is made, the change process may need such assistance as letting go of old ways, learning about alternatives, getting help from others, seizing opportunities as they arise—and, yes, faith and courage. All of these can be consciously processed.

Of course, self-efficacy develops with consciously driven skill acquisition and coping achievement. Even if or when we were not responsible for a given act at the time of the act, we are responsible for the original creation of the choice options that caused the act. This point is argued persuasively by Robert Kane,[72] who also quotes Aristotle as having said, "If a man is responsible for wicked acts that flow from his character, he must at some time in the past have been responsible for forming the wicked character from which these acts flow." Think about this the next time you feel as much empathy for a criminal as for his or her victim.

The mind is not limited to self-assembly as a stereotyped reflex response to experience. The conscious mind is assembled by the intervention of conscious thought in the response to experience. Thus, consciousness can empower or destroy people. With personal power, you no longer have to make excuses. The old habits can die away to be replaced by better habits created out of a free will to change.

> *We are what we repeatedly do.*
> *Excellence, then, is not an act, but a habit.*
> —Aristotle

Robotocism in Religion and Politics

Clearly, whether one is religious or has a certain political ideology is often influenced by how one was brought up as a child. Early childhood experiences "rub off" on us as more or less conditioned learning experiences. But how can anyone contend that all our impressions, beliefs, value systems, and preferences are not molded by conscious choice? Well, Nicholas Wade apparently makes this argument in his book *The Faith Instinct*. He argues plausibly that humans construct moral codes as a genetic imperative, since such codes have social benefit. Social animals in general benefit from genetic imperatives for rules that constrain self-interest in the interest of survival of the species. He says this moral *instinct* arises from an inherited brain-based moral system. This system operates subconsciously and is launched in the form of snap judgments when social situations call for an immediate response and there is no time for reflection and analysis.

Wade also asserts that humans have a consciousness-based moral reasoning system that, if there is enough time, operates like "a lawyer or public relations agent to rationalize the moral input it has been given [from the moral instinct system] and to justify an individual's actions to himself and his society."[73] Wade claims that the reasoning system means we are more than just genetically programmed robots. We are robots also sculpted by culture and learned values from schools, peers, parents, and religious organizations. Neither the instinct nor the reasoning systems, however, provide any room for free will. While I agree that human beliefs, moral and otherwise, are largely learned, Wade and others like him ignore the possibility that humans have choice. To them, being "born again" in any sense is not possible.

Another question is not so much why humans are creatures of belief but why the brain strives to create beliefs, religious or otherwise, in the absence of confirmatory evidence. For example, whether it is a belief in Islam or the future prospects of a stock, the brain will create beliefs based on certain undocumented assumptions.

Why does the brain do this? I think one reason is the filling-in-the-blanks phenomenon I illustrated with ambiguous figures in chapter 3. The conscious mind uses what little information it has to construct a more complete set of premises and an associated scenario, which it can then use as a framework for interpreting events, defining purposes, deciding what to do, promoting emotional well-being, and relating to others. Most of the time, operating under these assumptions works out pretty well and is not too destructive, even when the original assumptions are wrong.

Having hopefully made the case that science has not yet proved free will to be an illusion, I now make the claim that whatever our position on the matter is, it is a *belief*. What we believe in this regard does matter: it has enormous practical consequences, for it typically affects other beliefs, particularly regarding religion and politics. Other beliefs are affected because it is human nature to adopt belief systems that are compatible with one another. If we believe incompatible things about free will, we risk what psychologists call the distressing state of *cognitive dissonance*. Humans have an inherent motivational drive to minimize such dissonance. So belief about free will, illusory or not, needs to be applied consistently.

For a while, I participated in online discussions about free will, but I soon realized that the discussants believed free will to be an illusion. I eventually stopped participating because the others were extremely aggressive, ganging up on me in efforts to validate their belief system. That system included more than belief that free will was illusory. The other bloggers commonly volunteered the fact that they were atheists. Clearly they were using their blog posts to pursue an alternative agenda.

At first, I thought ideas about free will and atheism were unrelated, but now I realize it is all part of being consistent. The atheists I tried to interact with were on a mission, seemingly compelled to convince me of the rightness of their beliefs, much as evangelical Christians are. They contrived all sorts of specious assumptions and convoluted arguments to present their robotocist position. Notably, the atheists ignored all the freewill points I raised that they could not answer.

Free will is a big problem for atheists because a principal foundation of free-will belief comes from religion. Namely, people of faith generally assert that they are accountable to whatever god they believe in. It would not be fair to be held accountable if you have no free will. Robotocists need to reject the notion of sin, for it assumes that humans can control sinful behavior. If we can't control sin, then we are neither responsible nor accountable for it. Indeed, atheists may even cherry-pick the Bible, using the doctrine of "original sin" to prove their point that our misdeeds arise from the moral inadequacy that humans have because we are evolved animals. But some "sin," like evil, seems to be unique to humans. In all my years of working with animals, on the farm, as a veterinarian, and as an animal researcher, I have never seen an animal that was evil, and only a few that were malicious. I can't say that about people.

Unlike many other atheists, Nicholas Wade believes that religion is a good thing. He argues in his book that believing in free will and religion has social utility, even though both positions are wrong. He even argues that the capacity for so believing was commandeered in the evolutionary process because it had the advantage of promoting group loyalty, cooperation, and construction of social rules that promoted order and group success.

To the earlier point about cognitive dissonance in the context of politics, robotocists tend to think it is not fair that some people succeed in life while others fail, since the lack of free will means nobody should get credit for success (they have good genes, were lucky, or exploited others) and nobody should be blamed for failure (they have poor genes, were unlucky, or were the ones exploited). Thus, equal outcomes should be the goal of social and governmental reform.

Others obviously have the opposite and irreconcilable view. But to hold such a view, they are obliged to believe in free will. Or, to put it another way, because they believe in free will they believe people have personal responsibility for their station in life.

How can anyone seriously argue a person is not responsible for criminal and evil behavior? Yet this view of human life is endorsed by many highly educated lawyers, mental health professionals, and scientists. Oh, and don't forget certain politicians. These people demean human life by regarding us all either as victims or as undeserving beneficiaries of good fortune. Success can be demeaned because we don't deserve it. Failure can be excused because we can't help it. Since there are fewer successful people than unsuccessful ones, it is easy to understand why politicians pander to the unsuccessful.

Paternalism comes to be viewed as a necessary virtue of those who extend it. Ego gratification comes from being the "ruling elite." Those in charge see themselves as superior people, smarter, and more capable of knowing what is best for lesser beings.

This has enormous implications for the survival of freedom in a democracy. Historically, freedom erodes in a flood of tyranny when leaders decide who, if anybody, gets to be successful. The so-called liberating effect of the "Arab Spring" illustrates the seemingly inevitable fact that dictatorships typically get replaced by successor dictatorships. If you don't believe in free will, the whole idea of freedom becomes problematic, even irrelevant. Dictators can't allow it.

A new movement in US politics is the idea of "disproportionate impact." People who have not succeeded economically or in social standing are assumed by many on the political left to have been discriminated against. Government agencies are pressured to bend the rules to manipulate the disparity in impact. Thus, home loans end up being given to people who cannot make the mortgage payments, insurance companies are pressured to underwrite risky personal property and healthcare coverage, and employers are pressured to fill racial, ethnic, and gender quotas irrespective of job requirements. A person's success in life should be provided, even if it has not been earned. This view is often called the "fairness doctrine."[74]

If we believe there is no free will, how can we justify our criminal justice system? If people cannot make choices freely, and if all their decisions emanate from unconscious processes, then how can we hold them responsible for unacceptable behavior? All crime should be tolerated or at least excused because the criminal could not help it. The human robot committed the crime. This would mean that we should reform the criminal justice system so that no criminal is jailed or punished. If criminals can't stop themselves from bad acts, it is inhumane to punish criminals or even terrorists. The only justification for locking anybody up for misdeeds would be to protect society from further crime or terrorism. Capital punishment has to be banned, as indeed it is in many parts of the world.

Many defense lawyers increasingly use neuroscience inappropriately to convince jurors that the defendant was not responsible for his or her criminal behavior. They even have a name for this kind of defense: "diminished capacity." Lawyers are adept at stressing mitigating circumstances, arguing that criminal behavior is often caused by a terrible upbringing, poverty, social discrimination, bad genes, or brain injury. To be sure, most murderers have been found to have a standard profile that includes childhood abuse and some kind of neurological or psychiatric disorder. But many nonmurderers have a similar profile. How can lack of free will explain such difference? The reality is that most people have brains that can learn social norms and choose socially appropriate behavior. Ignoring those norms is a choice.

Bad brains can surely cause bad behavior. But it is equally true that bad behavior can cause bad brains. What you choose to experience, think, and do creates learning episodes, and all learning, good and bad, sculpts brain function and anatomy to shape what you will become. It is true that brain scans, for example, can sometimes predict that certain people will commit crimes or other antisocial behaviors. But nobody seems to consider the alternative to a "bad brain" cause of misbehavior: namely, that what people freely choose to think and do changes the brain in ways that make it more likely that similar thoughts and behavior will be repeated. People who voluntarily indulge mind-altering experiences, such as unsavory friends, drugs, or destructive ideologies and lifestyles, have nobody else to blame for those choices.

A most disturbing book, written by Laurence Tancredi, uncritically accepts the Libet-type research reports I described earlier.[75] Tancredi is a lawyer and practicing psychiatrist. Not surprisingly, the poster boy for his arguments was a psychopathic serial killer, Ricky Green, who was abused as a child and had relatives with serious mental problems. Thus, Tancredi stresses that bad genes and bad treatment as a child made Green become a "biologically driven" murderer. Yet, in recounting the case history, it became clear that Green was not insane. He was fully aware of his childhood past, and he was even remorseful about his murders. He was also aware that his out-of-control episodes were triggered by the combination of sex and alcohol. So it was clear, even to Green, that his crimes could have been prevented by avoiding alcohol. He apparently was not an alcoholic who had no control over his drinking.

Even if we give the benefit of doubt to the belief that Green could not control himself, it is a stretch to argue that the uncontrollability of psychopaths applies to everybody else. One would have to argue that normal people are only normal because they had the advantage of good genes and had a childhood in which their mental health was not damaged.

Interestingly, Tancredi acknowledges the contradictory fact that the brain is changeable if skilled therapists provide structured rehabilitation for dysfunctional thinking. But the general tenor of the argument is that the individual is powerless to contribute to such changes. It has to come from others. Because dysfunctional people are victims who seemingly can't help themselves, it is the duty of psychiatrists and government to mold the brains of people so they overcome bad genes and whatever bad experiences life has thrust upon them. People, since they are supposedly robots, do not have the power to nurture their own brains. Thus, government must create a cultural and educational environment in which humans are molded to conform to some predefined state of normality. Does this remind you of Aldous Huxley's *Brave New World*?

Tancredi and his crowd also do not believe that dysfunctional people might have become that way through their own freely determined bad choices along their life journeys. I argue that those bad choices may have even sculpted maladaptive changes in their brain function. Arguing that the brain is modifiable by experience is a two-edged sword. While one edge slashes the idea that a person can't change his or her brain, the other edge slashes the idea that people can't be changed by the influence of others or by their own conscious decisions.

A major function of consciousness, as I have argued, is to program the brain, which inevitably causes lasting changes in its structure and functions. If consciousness provides capability for freely chosen intentions and decisions, then people are responsible for how those powers are deployed.

Tancredi acknowledges that many people have bad genes and very traumatic childhoods yet overcome these disadvantages. Sexually abused children do not necessarily become sexual predators as adults, and they may, in fact, become crusaders to protect children from abuse. But they don't get any credit for a freely chosen decision to live a wholesome and helpful life. People who live a wholesome and constructive rarely receive any credit. Their virtue is attributed to necessity, not to anything they voluntarily chose to do. How then do we account for the effect of schools and religious teachings? Do we conclude that it is the inner robot that decides which ideas and beliefs to accept and which to reject? If so, why do some brains accept the teachings and others reject them?

The evidence for brain plasticity, for good or bad, as summarized in chapter 2 and elsewhere in this book, is overwhelming. Yet this evidence tends to get ignored when excuses are sought for inappropriate behavior. Similar excuse making is commonly applied to dysfunctional people who have had the misfortune of bad childhood upbringings. It is true, of course, that such children usually had no control over their circumstance, and that rising above them is surely hard, but they still have a human capacity to overcome as adults. Many thousands have done just that. To deny that people have such capacity is dehumanizing.

All of us have unconscious attitudes and drives that arise primarily from our animal-based biology. But our misdeeds should not be excused. How do we explain how so many people overcome their unconscious weaknesses, whether those are compulsions or addictions, maladaptive lifestyle, or even changes in attitudes and belief systems? How do we explain how brilliant people can do dumb things or how people of ordinary IQ sometimes achieve brilliance? At a minimum, our brain's destiny is guided by conscious programming of the unconscious mind. This power is exerted in many ways, ranging from deciding what we read and what we watch on television to the kinds of people we associate with and the kinds of behaviors we indulge.

Ethical behavior and personal responsibility are learned by the brain and subject to change by new learning. A terrible childhood, for example, certainly teaches the brain badly, but this need not condemn one to an immoral or underachieving life. Abraham Lincoln and Thomas Huxley are conspicuous examples of people who consciously chose to rise above their early bad environment. Sigmund Freud was a cocaine addict. George Patton hallucinated. Merriweather Lewis, of the Lewis and Clark expedition, was a manic depressive. True, some brains have abnormalities that make it much more difficult to learn the social construct or to obey it. Yet many brain abnormalities are created by the lifestyle and thought and behavioral choices that a person freely chooses. You will probably mess up your brain by snorting cocaine or smoking pot, but that behavior is something you choose to do. You may program your brain badly by associating with the wrong people, but again, that is a choice, not a necessity.

Let us leave aside for the moment the fact that a world in which no one can be held accountable is impractical, socially disruptive, and even intolerable. After all, these conclusions are not valid scientific arguments against the robotocist position. But hopefully my analysis of neuroscience research on free will shows that the robotocist view is poorly supported scientifically. In the next section, I will present arguments for free will based on behaviors of conscious avatars that are hard to explain with robotocism.

Rather than push their argument against free will, robotocists would do better to focus more on why people are so prone to deny culpability for their misdeeds, to rationalize, and to make excuses. The focus of my *Blame Game* was on showing people that they can freely choose to make better choices.

WHAT CONSCIOUSNESS DOES

The premise of natural selection and evolution is that body structures and functions generally survive the ravages of geological time because they are useful for survival. Consciousness evolved and survived. Of what use is consciousness if all it can do is observe? Indeed, if it is useless, consciousness might lead to despair if it could not do something to confront challenge and adversity. On the other hand, if our consciousness is only an observer, it would be imperative at least to *believe* that our conscious can keep us safe from helplessness. If free will is an illusion, it is certainly a useful one.

In any case, free will or not, we can know consciously when we feel defensive or angry or overly critical or judgmental or upset. We can know when we try to make excuses. We can also recognize the situational causes or triggers of undesirable attitudes and behavior. Further, we can realize that avoiding the situational causes or triggers will not always work, that real change must come from within. Is such knowledge not useful?

Many scientists argue that consciousness can't do anything, that all willed action is generated unconsciously. But their arguments are confused. For example, Zoltan Torey says that the conscious executive function becomes active when the unconscious mind senses the options (as in smoke or quit smoking).[76] While still maintaining that conscious mind is not part of the decision making, Torey, in seeming contradiction, in the same paragraph concedes that conscious mind creates the "values and valences that are the guidelines for behavioral decisions." He goes on to assert that "this autonomy from within is more than an immense achievement. It is the only form of freedom that a self-enclosed monistic universe can lawfully generate."

Even if the human brain lacks free will, that doesn't prevent consciousness from being engaged in implementing instructions from the unconscious mind. Even the advocates of illusory free will, including Libet himself, grudgingly concede the point that conscious mind can veto unconsciously made decisions. An everyday example is provided by cigarette smokers who, as their blood nicotine level declines after the preceding smoke, feel an overpowering unconscious urge to light up once more. The avatar, however, if fully convinced from research reports of a link between smoking and lung cancer, can say "stop it!" and prevail. Really, the argument is not about free will but about whether conscious mind can generate executive functions, even if it is instructed by unconscious processing.

One could challenge this view by demanding some evidence that there are things a waking person can't do if they were not conscious. That is a stacked-deck question because it is nearly impossible to have humans who are awake but not conscious. These two things typically go together. However, in the change-blindness conditions described in the next section, titled "Directing Attention," people are awake but cannot respond to what they see. Perhaps it might be useful to conduct some studies with transcranial magnetic stimulation, which reversibly blocks certain areas of neocortex and associated conscious function, while testing to see if a waking person can perform any actions that ordinarily require that part of cortex to be functioning.

I have four reasons to contend that conscious mind does things:

1. There are many brain functions, including doing what is needed to stay alive, that cannot be performed when unconsciousness.
2. The quality of executive functions is proportional to changes across

the spectrum of consciousness, from drowsiness to intense arousal and
attentiveness.
3. Both consciousness and executive functions arise from CIPs in the neocortex. They share the same neurons and cortical circuitry.
4. CIPs of nonconscious and unconscious mind certainly do things. It
seems reasonable to think that CIPs of consciousness can do things.

Critics will still insist that executive action comes from unconscious mind,
that free will is an illusion, as is all conscious will. The evidence for the former is
flawed and unconvincing, and for the latter the evidence is absent.

Each person's avatar usually tries to take care of itself. Indeed, that is where
humans get their innate drive to live and their fear of death. As explained earlier
in the consideration of agency, consciousness is our brain's "executive agent."
Lower mind levels can only prod the avatar in terms of innate drives and motivations, many of which have been learned, either from inadvertent experience or
from deliberate choice and arrangement by the avatar.

As my own avatar, I have the liberty and power to decide what is best for
my brain and body. I can train my brain or I can be controlled by it. If I don't
like the things my lower mind wants me to do, I can veto those "suggestions" or
even compulsions. I seem to have some free will that empowers me to weigh the
pros and cons of various actions and decide what evidence-based action would
best serve my best interests. I even get to decide what my best interests are. Some
people are really good at self-discipline. Military boot camps prove that unruly
teenagers can learn self-discipline.

In the case of the smoking example, unconscious mind presents the options
of light up or quit smoking. Conscious mind then assigns priority scores based
on what it has learned about the pleasures of smoking versus the likely health
hazards. But what mind decided whether to read about or believe what is written
about the health hazards of smoking? What mind decides "I need to see more
data" or "I have seen enough; my mind is made up." In short, I contend that the
avatar programs unconscious mind by controlling what is selected for learning
and what belief position is taken.

So I feel compelled to assert that I am responsible for who I am and who I
may become. Of course, I am constrained by the limits of my lower minds and

their biology. The longer I live the more I realize that I and all other avatars can be captains of their ship, masters of their soul. Note I said "can be." We avatars don't always believe and act on that premise. Indeed, those avatars who doubt their power don't exert it, and they then become victims of their own limited belief system. Of course, captains sometimes run their ships aground, and we avatars often do the same thing when we are not paying attention or making the right choices. Consciousness is always being pushed and shoved around by what the unconscious mind wants to do.

Directing Attention

Many readers have seen the video (currently also on YouTube) in which a group of people are passing a basketball around.[77] They are instructed to count how many times the people in white clothes pass the basketball in any direction to any person. Somewhere near the middle of the video, a person in a gorilla suit walks through the group and leaves. When viewers were asked what they had seen, most failed to report seeing the gorilla.

Actually, they did see the gorilla. That is, their eyes saw it, and no doubt it was detected along the lateral geniculate part of their thalamus and the visual cortex pathway. But it was not perceived consciously. Such phenomena illustrate what is known as "change blindness." That is, under certain conditions, like the gorilla example, a visual scene change occurs and is seen with the eyes but is not consciously detected. But change is a subset of a more general phenomenon of selective attention.

Detecting and perceiving are distinctly different. Perceiving requires consciousness. The eye detects, but the conscious mind *perceives*. A similar distinction can be found with other senses, including pain.

In the case of pain, I learned early in my training as a veterinarian to distract a cow's attention from a needle injection by slapping its leg as I simultaneously thrust in the needle. Dentists do the same thing when they twist your cheek as they insert the needle for the nerve-block injection. In both cases, the pain is not noticed because of a focus on other stimuli.

Such phenomena are important for understanding what consciousness is and how it works. Let us think about this the way a neuroscientist might.

First, it is apparent that change blindness occurs because the conscious mind is not attending to all simultaneous events. Indeed, consciousness can't attend to everything because it has limited information-carrying capacity.

What does that tell us about attending? It seems to me that conscious attending is the mechanism by which things can get consciously registered. Do the details of the CIP representation of a stimulus change depending on whether or not it is seen unconsciously or perceived consciously? It probably does, and I have suggested as much to fellow neuroscientists, suggesting that they look for such differences. They haven't done it yet.

Another element of this change blindness phenomenon is the matter of expectation. In the example above, if the conscious mind knew in advance that the scene would include a gorilla, the gorilla would most likely have been perceived. In other words, we have the not surprising conclusion that we see what we expect or want to see. That is related to the well-known phenomena of "hearing what we want to hear" or "believing what we want to believe."

This brings us to the role of consciousness in intention. If we intend to accomplish a certain kind of task, such as seeing the gorilla when it appears, we are much more likely to make it so. We should think of intention as being critically linked to achieving focused attention. Intending to pay attention to a specific thing is a special capability of consciousness.

Recall our earlier discussion about perception of ambiguous figures. Most ambiguous figures can be perceived in one of two ways, as in figure 3.3, a drawing that can look like either a man's face or a naked lady, depending on how you think about the stimulus. Some images have three or more alternative percepts. Regardless of how many alternatives there are for a given image, the principles of how the brain handles the information should apply equally.

We found in our studies that each person had a default percept for each kind of image. For example, when looking at the face/body image, a given person could readily see the face but have to think hard about finding an alternative image, which would eventually "pop up" as a female body. The default was not the same for each person. Some people defaulted to the face image, for example, while the body percept was the default for others.

What does this tell us? It may reflect how well-learned a given percept was. If my default percept is the face, for some reason my mind has over-learned what a face looks like compared to a naked female. (You don't need to know what my default is). So the first thing that pops into consciousness is the best-learned percept. As with all forms of memory, if something is not well learned, cues are needed to help recall it. In the case of the illusions, we found that many subjects had difficulty seeing the alternative percept of a given figure. They had been told that every image had an alternative percept and that their task was to detect both. For some images, they could only perceive one of the alternatives. For the other alternative, they needed hints. For example, in the face/female body image, the experimenter might tell a subject whose default was the naked lady, "Focus on the two v's." If that didn't trigger a face recall, we might have to cue further by saying "think of the two v's as eyes." For subjects that didn't see the female, we might cue with "think of the curved line next to what you think is a nose." If that was insufficient, we would add, "think of that curved line as a breast." Once such cues were given, odds were good that associated features would serve as innate cues to make the whole percept become apparent.

Ambiguous figure perception is not quite the same as change blindness because the image does not actually change. The conscious mind perceives one of the alternative images at a time but can switch percepts even though the stimulus pattern on the retina is constant. The selective nature of the attention is evident from the fact that you can only perceive one alternative image at a time. Moreover, typically you have to selectively attend to certain elements of the image in order to perceive the alternate image.

Why is attentiveness capacity limited? Attending must consume circuit resources, just as multitasking does. This is consistent with the idea that conscious working memory has very limited capacity. Think about it: just to generate a conscious mind in the first place must require a great deal of circuit resources. That is why only human brains are good at producing robust consciousness. Only humans have enough extra neurons and circuits that are not needed just to sustain robotic behavior. There may not be enough circuit resources left over to handle much information in conscious working memory.

From watching the multitasking abilities of today's young people, you might think that the brain has enormous conscious carrying capacity. But, as we discussed earlier, evidence from multitasking research shows that the brain can only do one thing at a time. These studies indicate that the brain is just rapidly switching among parallel circuits at such high speeds that it appears to be simultaneous.

Actually, this is not all that odd. Computer programs work in a similar way. When you are writing to a document in one program and downloading a file in a browser program, both processes seem to go on at the same time, but actually the computer is switching at high speeds between the two operations.

Why the need for switching? In both computers and brains, the reason may be a circuit limitation. One circuit may not have enough processing elements to do everything at the same time. Another possibility is that the processes operate at the same time, but the recognition or readout of the result has to occur stepwise. If, for example, you tried to display both alternatives of an ambiguous figure at the same time in the "mind's eye" the result would be a morphed jumble of male face and female body that would be unintelligible. In other words, your brain reads out one image at a time because it is not useful to see confusing morphed images. A brain built to go through life confusing itself is not a very useful brain.

So now we can see that what the conscious mind perceives depends of several things: (1) expectations, (2) selective attention, (3) memory stores, and (4) associational cues. This brings us to the role of consciousness in learning.

Learning

Learning can of course occur unconsciously, as amply demonstrated in classical and operant conditioning. Classical conditioning was demonstrated by the famous Russian, Ivan Pavlov, who showed that dogs could learn an association between a bell and subsequent presentation of food. A dog so trained will start salivating and secreting gastric juices when it hears the bell. Operant conditioning is illustrated by common training techniques for circus and work animals. Complicated behaviors can be shaped by small successive steps in which rewards are repeatedly given when an animal accidentally does something that approximates what you want it to do. Gradually, the ante is upped so that the animal only gets rewarded when a behavior emerges that is a better approximation of what you want the animal to do.

All three minds can learn, and the nonconscious and unconscious minds are particularly good at conditioning. Nonconscious mind, for example, can create supersensitive reflexes, especially under certain pathological conditions. Unconscious learning results from unconscious registration of cues and associations, especially those affecting bodily functions or emotions. Persistent high blood pressure in the absence of artery obstruction can be an example. Likewise, we can learn things unconsciously, such as bias and emotions.

The unconscious mind uses its store of implicit memories to influence our attitudes, emotions, beliefs, choices, and behavior—both conscious and unconscious. The process is reciprocal: our behavior, good, bad, or otherwise, creates informational feedback and learning experiences that get deposited in the unconscious and thus become a part of us.

For example, abused young children may not have any conscious memory of the abuse. But those memories may still be etched in places like the amygdala and other emotional parts of the brain. Though not normally accessible to the conscious mind, these emotional parts of the brain can still be active in the unconscious mind, and they can influence how we feel and act beyond our awareness.

Unconscious memories and thinking routinely get expressed as "subliteral" meanings in human communication. That is, much of what we say has an obvious literal meaning and a less obvious, and sometimes very different, unconscious meaning. This idea applies to body language, of course, but it easily extends to encoded talk. The idea is related to euphemism, "political correctness," "doubletalk," and of course, "reading between the lines."

The conscious mind learns at its own level, but it also expedites the unconscious learning process, which once accomplished allows the brain to use its learning more efficiently with less effort. This is illustrated by such common experiences as learning to tie shoelaces, ride a bike, play a musical instrument or skill sport, and so on. Conscious thought identifies what we want to learn, motivates us to learn it, facilitates association and integration into our current conscious understanding, and implements specific learning strategies, tactics, and corrective feedback.

We can learn from our mistakes in an unconscious, Pavlovian manner or we can use conscious analysis to make such learning more effective. Moreover, consciousness can enhance learning in other ways, such as learning how to correct mistakes or how to repeat successes. I am reminded of an anecdote I once heard attributed to T. Boone Pickens, the famous billionaire who has won and lost several fortunes. As the story goes, when he was asked about the secret of his success, Pickens explained that a beloved basketball coach once told the members of his team to learn from their mistakes, but don't dwell on them. "Instead," he said, "dwell on what you did right and do more of that!"

This coach might have taken psychology in college and learned the power of operant conditioning, which is readily produced in lower animals with little or no capacity for conscious analysis. In training a parrot to perform a series of acts, you don't punish him when he does something wrong. You reward him when he does something right. When this basic principle of learning is fortified by conscious direction, learning becomes much more robust and lasting.

The programming function of consciousness arises from a better ability to pay attention to things that are important, such as in the example above; that is, focusing on doing what works. Consciousness is a major reason why people have such a powerful capacity to learn. Consider how impossible it would be to learn a language or play a musical instrument if you had to do it unconsciously. In the case of language, I could unconsciously learn by much-repeated Pavlovian association the conjugation of a verb. But through conscious thought, my avatar can memorize it in a few seconds. There is, of course, unconscious language, preserved in memory and demonstrable by word-priming experiments. But such language capability is not fully operational until it resides in conscious mind. Unconscious mind might generate the will to learn a specific thing, but it lacks the explicit information on how to do so until after it has been learned.

Like the ability to ride a bicycle, much of what we have made of ourselves from conscious thinking and behavior has been driven underground into our unconscious as it becomes well learned. Remember when you first learned to ride a bicycle? You surely could not have done that without conscious awareness of the situation and the bodily actions that were necessary. This is true for virtually all complex movement tasks, as in learning to type, play sports, play music, draw, write cursive, and so on. In my blog posts I have repeatedly stressed the developmental benefits of promoting sports, music, cursive writing, and the like for children. They not only learn the skills, they improve their ability to learn in general.

Consciousness is especially crucial in early learning stages. When you learn a golf swing, for example, you have to consciously think about how to grip the club and how to do the back swing; you have to remind yourself to keep your eye on the ball; and then you have to remember to make a full follow through. Even simply reaching for an object may require conscious control if conditions vary. For example, in a study of monkeys from which nerve impulses were recorded from the premotor cortex, the impulse patterns varied systematically when obstacles were placed in different ways in the path of their reaching movement.[78] In other words, these neurons were capturing spatial information that the conscious monkey was using for planning strategies to accomplish the movement.

Explicitness Promotes Memory

If the conscious mind does nothing, how do we explain its role in memory formation? Common experience establishes that focused attention, rehearsal, and deliberate use of mnemonic devices all promote memory consolidation. These things are done consciously. True, many memories if sufficiently rehearsed in consciousness become stored as unconscious procedural memories, as evident in the formation of habits, prejudices, and motor memories like touch typing, riding a bicycle, and sports skills. But none of these habits develop without prior repeated conscious processing.

One major function of consciousness has to do with efficiency of working memory, the limited-capacity conscious memory that we use, for example, when looking up a phone number and remembering it long enough to dial the number. Consider the likely possibility that consciousness uses its working memory as a virtual notepad for: (1) selection of choices and agency, (2) long-range planning, (3) construction and storage of memories, (4) retrieval of stored unconscious memories for use and modification, and (5) "troubleshooting" and reflective analysis and decision making.[79]

Representations of sensations and ideas are held in working memory and provide a necessary support for orderly conscious thinking, as shown in figure 3.4. It seems likely that working memory operates both unconsciously and consciously, but conscious working memory can be sustained indefinitely and, of course, after sufficient rehearsal, the memory can become permanent.

I want to emphasize that conscious working memory can enhance the quality of thought because the thought is sustained explicitly. Moreover, since memory consolidation is a function of how long and well representations are rehearsed in conscious working memory, consciousness is a major contributor to the human capacity to form lasting memory.

Consciousness is the brain's master teacher, directing the programming of unconscious operations. We consciously decide what to learn by choosing what to read, what to listen to, what movies and television to watch. We can consciously decide which behaviors to engage in and which to avoid. We can consciously determine what we will think about and not think about. We also consciously generate intentions, decisions and plans—the essence of free will.

Introspection

The great thing about consciousness is that it enables introspection. Humans can explicitly think about what they have done, are currently thinking, or will do. This capability is the precursor to the brain's ability to program itself. Through introspection, we can test whether our thoughts, beliefs, and attitudes are appropriate and adaptive, and in the process we can explicitly determine what if any changes need to be made. We can think "What did I do wrong? How do I prevent this from happening again? What did I do right? How can I replicate this behavior?"

Consciousness allows us to comprehend the world and ourselves more specifically and more accurately. Psychotherapy is based on the premise that consciousness helps us to know ourselves. In most normal people, introspection occurs without the need for much outside assistance. The consequences of not knowing ourselves are enormous. If we do not know and face our fears, for example, we have little chance of conquering them. If we do not recognize bad habits, how will we correct them? If we do not know when our thinking is irrational, how can we learn to think better? If we do not know when our emotions are inappropriate, how can we learn to deal appropriately with the disturbing events of life? If we do not know what kind of person we are and how we ought to improve, how can we grow? If we don't know we are making excuses, how can we stop doing so and address the real problems?

Robotocists, to be consistent in their argument, would have to claim that explicit introspection occurs unconsciously and that consciousness is then informed. Even if conscious introspection and analysis can only facilitate unconscious thinking, consciousness is still doing something.

When we use consciousness's capacity for introspection and analysis, we increase the odds of spotting flawed attitudes and thinking as well as the options for better alternatives. Unfortunately, the conscious mind's natural tendency is to believe its initial conclusions and not to second guess them. In short, a main value of consciousness is that it has the capacity to challenge its own thinking. Doesn't willingness to do so require some free will?

Language

Logical thought is greatly facilitated by good language skills because language is the medium that carries much of our explicit thinking. Mathematical thinking is similarly enhanced by good math skills. Deductions can be made unconsciously, but consciousness allows us to make explicit the premises and propositions from which we derive conclusions. Inductive thought may begin in the unconscious mind, but consciousness allows us to make explicit the number of particulars or specific instances from which we can create a generalized conclusion.

Of course, language is not the only venue for conscious expression. Consider art or music. We consciously conceive and interpret images and sounds. We can also be aware of our emotions and the varied input of our senses even when it is difficult to explain these things in words. There is also body language.

Consciousness, especially when it recruits language, gives us more effective interpretative and analytical capacity. Many of us think through issues and problems with conscious use of silent-language self-talk. Other people use imagery for creative problem solving—Einstein was a good case in point, as for example his visions of riding on a beam of light or train motion past a station. In all such cases, the common denominator is consciousness.

Reasoning

Consciousness enhances reason. When people do really stupid things, we accuse them of being "mindless." In other words, such actions occur when people don't consciously think about the why, what, and how of their thinking. These things constitute reasoning, which occurs best in consciousness.

Animals can seemingly make logical decisions, but they cannot sustain a long reasoning process that involves multiple steps. Their reasoning capacity is limited because they do not have a level of consciousness that allows them to shuttle onto and off of their working memory "notepad" (recall figure 3.4). This point is illustrated by what a teacher of students with learning disabilities told me. She said, "My students can solve the same math problems that normal students solve, but they can't consciously remember the sequence of steps."

Despite all these powers of consciousness, the problem is that human consciousness is not always up to the tasks imposed by the complexities of life.

> *Surely, I am too stupid to be human.*
> Proverbs 30:2

Yet consciousness is the hallmark of being human. It gives us the ability to recognize issues, problems, opportunities that involve ourselves and others. But we are too stupid to cope completely, to think as effectively as we ought.

Limited though it is, the power of conscious reasoning sets humans apart from all other life forms. We not only think; we have the too-often-unused capacity to think logically. Logical thought, as noted earlier, is greatly facilitated by good language skills because language is the medium that carries much of our thinking in explicit ways. Not everybody has good language skills. Glaring examples of flawed thinking, even by scholars, are provided in Nassim Taleb's book *The Black Swan*.

Deductions, intentions, and decisions can be made unconsciously. But consciousness allows us to make explicit the underlying premises and propositions, and it allows us to hold them long enough in conscious working memory. Inductive thought likewise begins in the unconscious, but consciousness again allows us to make explicit the number of particulars or specific instances and alternatives from which we can create insights and a generalized conclusion.

Consciousness Promotes Better Decisions

The capacity for conscious choice gives us more power to make better decisions than our robot unconscious mind. Consciousness allows us to explicitly weigh the pros and cons of multiple alternatives and to choose from among them. Consciousness allows us to monitor the anticipated consequence of decisions and to choose to make necessary adjustments. Many robotocists like to point to automatic behaviors, such as driving, riding a bicycle, and the like, as good examples of the absence of conscious control. This position, as I mentioned before, does not recognize that it was conscious thought that caused these automatic behaviors to be learned well enough to become automated. Delivering the instruction and initially learning it was hardly an unconscious process.

The conscious mind gives the brain a chance to get a second opinion. The conscious mind provides a second level of analysis and judgment. Consciousness is like having an editor for your manuscript for living. It doesn't matter whether the initial drafts of our manuscripts come from the conscious or the unconscious mind. What does matter is that consciousness can review what we have been thinking and edit the script.

Even when decisions are made unconsciously, the conscious mind can veto or adjust decisions that it concludes are not in the best interests of the body or brain. Now the question is, how is the second-order decision made? Vetoed acts are commonly decided in the conscious mind. Are such vetoes also constructed unconsciously such that the conscious mind just appropriates the veto or adjustment as its own after the fact? This may be hard to determine, since in many real-world situations control over implementation of decisions is made in fractions of a second, often with no chance for prolonged unconscious analysis, planning, implementation of a response, and then ancillary transfer of all that information to consciousness. As I write, having just watched the 2013 Wimbledon tennis finals, nobody can convince me that Andy Murray and Novak Djokovic played like robots, except during transitory mental lapses. Any tennis player or fan knows that when you do play like a robot you make many unforced errors. Tennis players become champions only when they have learned to sustain intense conscious focus and positive emotion throughout a match.

Not all of us are equal in terms of the ability to exert conscious executive control. Some people are more autonomous and self-actualized than others. Some people are servile and submissive, more comfortable in choosing to blindly accept the views of gurus, teachers, or political leaders. It is not clear how much of this inequality is due to genetics and how much to learning, but certainly we can learn such aspects of control over unconscious mind as how to control impulses, how to make wiser decisions, and how to plan carefully. You can call this "growing up." Like any other kind of willful learning, the early stages are likely to be dominated by freely chosen conscious analysis, choice, and self-directed guidance.

Consciousness has obvious brain-programming functions. Explicit thought provides for deciding what to believe, how to think of ourselves and others, how to perform novel and complex tasks (such as learning to read, type, play music), and how to plan for the future. Consciousness does more than simply program the brain. It helps the brain set the parameters by which it will operate and be programmed by actions in the future. In his book *The Neural Basis of Free Will: Criterial Causation* Peter Tse extends the role of conscious free will to possible future decisions. He makes the salient point that a conscious choice does not necessarily depend on what happens at the instant of the choice.[80] CIPs not only carry messages in the present, they also create action preferences for future choices by modifying the synaptic weightings within those circuits. Thus, a given circuit may operate differently in the future than it does in the present, even under the same decision-making conditions. For instance, conscious thought can perform explicit analyses of several anticipated future decisions that ultimately determine criteria for which choice will be made more or less automatically in the future. Or, more generally, conscious thought can generate schemas that are engaged in future choice situations. The explicitness of conscious thought provides special advantages for decision making, not only in the present, but also in response to future choice contingencies.

Creativity

Not only does the brain contain CIP representations of things we have experienced, but it also can create CIP representations of things and events that we have never experienced. Creativity is a marvelous mystery that no one really understands. Clearly, creating a representation of things that have never been seen or experienced requires combining in unique ways the CIP representations of things we have seen or experienced. No one knows how the brain decides which circuits to tap into to create original thought. No one knows why some brains are better at the creative process than others. Nor do we know much about how brains can be taught to be more creative. Few understand or even ask why dream content is more creative than the thinking we can usually generate during wakefulness. I suspect that dream-content construction has the creative advantage of not being constrained or "corrected" by ongoing feedback from events in the "outside" world.

Ideas that emerge in a dream are sometimes used as an example of unconscious thinking, but remember that dreaming is in itself a special kind of consciousness. The dreaming brain makes its thoughts explicit and thus accessible for subsequent conscious thinking operations in the dream and later during remembering of the dream.

Creative thinking also springs forth from conscious processes. Creative ideas are surely refined by conscious analysis, and they may also emerge from the process of striving for insight. Creativity is promoted when we consciously decide we want new ideas and are open to "thinking outside the box." Conscious processes structure our mental environment to facilitate creative ideas. We may decide to go on vacation or change the nature of our work, for example. As a result, fresh ideas often emerge. Or we may brainstorm, singly or in a group, in an effort to consciously identify numerous relevant possibilities, and the mental associations of these thoughts often trigger new ideas. Consciously searching for all reasonable alternatives greatly expands the range of possibilities that can contribute to creative thought.

Conversely, creative thinking becomes obstructed when we have conditioned our unconscious minds to mindless conformity. Much of such conformity occurs consciously. Knuckling under to peer pressure is a common example. Many people decide not to be original thinkers because they fear being labeled an "egghead," or "radical," or "out of step," or "not one of us." A major problem in schools is that many students feel it is not cool to be smart. This is one of many examples where consciousness betrays our best interests.

Reprogramming Conscious-Self Dysfunction

Consciousness provides a way to modify the sense of self. We can, through learned experience and self-talk, change our perceptions and representations about ourselves. Modern-day philosopher Patricia Churchland argues that particular self-representations can be spared when others are impaired.[81] Some amnesic subjects, for instance, have no conscious awareness of things that happen outside a small window of a minute or so of current and recent time. Such people are self-aware only in such near real time. For them, self-representation of the past is not possible, because the parts of the brain that form memory have been damaged.

A different example of self-dysfunction is schizophrenia. Schizophrenic patients have good autobiographical memory, but they are deeply confused about the boundary between self and non-self. For example, a schizophrenic may respond to touch stimulation by claiming that the sensation belongs to someone else or that it exists somewhere outside the body. "Voices" that schizophrenics hear are apparently silent self-talk that is incorrectly represented as coming from the outside. Unfortunately, it seems that such hallucinations cannot be cured by learning. That is, you can teach schizophrenics that they experience delusions, but that doesn't stop the hearing of voices and so on.

Each person's mind evolves in real time because each person's brain evolves in real time. The human brain evolved over millions of years as prehuman and even human brains generally got bigger, but it also evolves in each person as the brain circuitry is sculptured by maturation and experience from the womb to the tomb. Teenagers do not have the same brain they had before puberty, nor are they the same people they were before puberty. Some teenagers change almost overnight from a lovable cherub into a rebellious monster. You are not the same person you were as a teenager, and you do not have the same brain you did then. "Thank goodness," you probably say. But you also do not have the same brain you had ten years ago—or even yesterday! We all learn things every day, and learning inevitably makes physical and chemical changes in the brain as the new learning is stored.

The brain is maturing in very observable ways up to at least the age of twenty-five. For example, fiber tracts and their connectivity still progress up to that time. Microscopic changes in synapses and their biochemical machinery occur throughout life. What this means is that what we experience and learn actually changes the brain. Those changes influence changes in our mind.

Even the brain's topographical maps can be individually sculpted by unique experience. James Shreeve has summarized some of the studies that show the brain to be very changeable by experience.[82] For example, blind people who read braille show a great increase in the size of the region of the cortex that innervates the right index finger, the finger used to read braille. Violin players have an analogous spread of the cortical region that is associated with the fingers of the left hand. London taxi drivers have an enlarged rear portion of the hippocampus, a brain area involved in spatial orientation. Learning how to juggle increases the amount of gray matter in two cortical areas involved in vision and movement control. When newly trained jugglers stop practicing, these areas shrink back toward normal size.

Personal Growth

We can set goals and make plans to achieve those goals. Consciousness tells us what we need to learn to achieve those goals. Common experience, however, teaches that personal growth depends on conscious recognition of personal weakness and opportunities for self-improvement. Conscious intent and planning can guide us in making changes from within.

Brain activity causes intentions, decisions, and assorted behaviors. Consciousness arises from and is part of such brain activity. Therefore, the brain activity of consciousness should be able to cause and modify other brain activity. Personal consciousness facilitates two main functions: (1) it provides a personally relevant high level of thought that can explicitly integrate past, present, and future, and (2) it programs the unconscious mind's way of understanding and responding to the external world.

All of this ties to the fundamentals of self-identity. Our self-awareness of ourselves is in a constant state of flux. We transform who we think we are by how we think of ourselves. The enlightened brain nurtures this self-image, using consciousness to monitor and adjust the brain's engagement with the world, hopefully in the most adaptive ways. Further, how we think of ourselves continuously programs the brain, transforming who we really are, which in turn can change the way we think of ourselves and behave. In other words, how conscious mind thinks of itself becomes a self-fulfilling prophecy. This is the universal theme of motivational speakers.

Consciousness gives us the capacity to know ourselves, and to change what we don't like about ourselves. Personality growth depends on recognition of personal weaknesses in need of change. This was the main point of my *Blame Game* book.[83] We can know when we feel defensive or angry or overly critical or judgmental or upset—and even when we make excuses. We can also recognize the situational causes or triggers of undesirable attitudes and behavior. Further, we can realize that avoiding the situational causes or triggers will not always work, that real change must come from within. Conscious intent and planning can guide us in making those changes from within.

Personal Responsibility

Exercise of free will does not depend on whether you believe that conscious mind is an external "observer" in the head, as René Descartes did, or whether you think of consciousness as an avatar emerging from widely distributed coherent neuronal impulse activity in the neocortex, as I do. The important practical point is that consciousness can change the processes that generate it. This is a profound property that affects how our individual personalities and attitudes change over time, and it even affects the evolution of whole human cultures.

To believe in conscious choice is to take responsibility for one's actions and to hold others accountable for theirs. Moreover, taking responsibility leads to more success in life, however one wishes to define "success." Taking responsibility empowers people, enabling them to cope and overcome.

Whatever the nature of that conscious "ghost in the machine," it has the power to tell the brain what to think and do. The conscious mind can do more than just veto or regulate behavior on the fly. It can program the unconscious mind to generate vetoes under conditions in which the brain has learned not to do certain things.

The conscious mind is our primary agent of change. If we are ever to move up out of our comfort zone, it will be willed intent of the conscious mind that makes us do it. The conscious mind makes choices in the full light of awareness.

The unconscious mind makes choices, too, but the conscious mind is not aware of what the unconscious choices are until those consequences become explicit.

Having hopefully made the case that the conscious mind makes choices, and maybe even does so freely, I am especially obliged to explain why so many people make stupid and irrational choices. Read any day's newspaper and you will find examples of bad decisions. People often choose to believe in ideas that are demonstrably wrong or irrational, even to the point of suicide murders by Muslim jihadists, for example. Indeed, whole cultures have belief systems that are logically incompatible with the belief systems of other cultures. Clearly, somebody gets it wrong.

The explanation for bad choices involves a host of variables involving biology, personal experience, and, of course, conscious choice. Any particular brain can evolve any particular mind, depending on what the brain has learned and the choices that brain makes and has made. Societies, through education, politics, or religion, can influence conscious choices and use peer pressure or actual force to achieve compliance with those choices.

One thing is clear: our choices make us who we are. Few have put it any better than J. K. Rowling's fictional headmaster Albus Dumbledore, of Harry Potter's school of wizardry and witchcraft, in *Harry Potter and the Chamber of Secrets*:

It's our choices, Harry, that show what we truly are, far more than our abilities.

Belief

In his recent book *The Biology of Belief*,[84] Bruce Lipton argues that beliefs control our biology. His point is overstated, but from what I will say shortly about epigenetics, you should accept that his core idea has merit. The conscious mind does exert enormous control over the body. In medicine, this is evident in the placebo effect, by which subjects in clinical drug trials that are given a fake drug, a placebo, will also show improvement simply because they believe that they are getting the drug. Placebo effects are especially noticeable in the treatment of asthma, Parkinson's disease, and depression. Of course, if the drug being tested is any good, the treatment group shows a bigger effect. One cannot explain away the placebo effect by saying that the improvement is "all in their mind." It is more than that. The mind's belief has actually changed the bodily function. Many experiments document how the mind influences our heart rate, our blood pressure, our release of hormones, our immune system, and many other functions.

The opposite of the placebo effect, the *nocebo* effect is perhaps more common. Here the idea is that negative thoughts actually promote or aggravate medical problems. Psychosomatic diseases are the classical example. Typically, the magnitude of placebo and nocebo effects is about the same. Both phenomena create practical problems in research and development of new drugs. In drug testing, volunteers have to be informed of possible side effects, which negative thinkers come to expect will happen to them.

One thing that both placebo and nocebo effects demonstrate is that the conscious mind does have a way, albeit indirect, to influence nonconscious mind. Scientists know that some of this is mediated through the autonomic nervous system headquartered in the brainstem, but this doesn't seem to explain everything.

The point, for our purposes, is that what we believe is a matter of conscious choice. We can choose what we believe and expect. We can choose to embrace positive thoughts or negative thoughts. We choose whether to see the glass as half full or half empty. We can choose religious beliefs that lift the human spirit or beliefs focused on killing those who do not share our religion. Of course, in some countries, you are under great religious or political pressure to believe only certain things.

Bruce Lipton likens the unconscious to running the mind on autopilot, whereas conscious mind provides manual control. The most obvious examples of how manual control of conscious thinking serves to reprogram the unconscious mind are the ways psychologists treat phobias. We may, for example, be unconsciously driven by fears of snakes, or heights, or of failure, or even of success. Careful analysis of these phobias in the light of conscious and rational analysis can reprogram the unconscious "autopilot" so that we no longer have such fears. More generally, we can say, as Lipton does, that "the biggest impediments to realizing the successes of which we dream are the limitations programmed into the unconscious."

Beliefs that are embraced in the unconscious are automated and rise to the surface rapidly and without analysis. This is the mechanism of bias and prejudice. Such beliefs often serve us poorly.

The conscious mind has the option of considered evaluation of beliefs, and it can choose which beliefs will serve us well and which are counterproductive. Thus, we rely on the manual control of the conscious mind to superimpose its will and belief systems on our thinking. Ultimately, the belief systems that become embraced by the conscious mind can serve to reprogram our unconscious mind. This even occurs at the social level. In our own time, witness how attitudes about racial segregation have caused most people to realize that it was unfair, and most have adjusted their attitudes accordingly.

The issue about free will is not so much whether we have any, but whether we act on the belief that we do. Philosopher Immanuel Kant argued that we are essentially free and fail to live up to the promise of such freedom if we believe that biology and external forces rule us. Existential philosopher Jean-Paul Sartre, despite all the fallacies of his irreligious and political beliefs, had it right with his arguments that humans do not have a fixed nature handed out from biology. Humans make their own nature out of the freedom to do so in the environment in which they exist. Humans are regarded, in existential philosophy, as independently acting and responsible conscious beings. Each individual brain is uniquely constructed from life experience, and much of that is self-constructed by willful choices. Science does not hold that the avatars of the human brain are independent of genetics or environmental programming. Rather, each avatar is an independent being, acting in the world as a distinct entity, not as some Borg-like unit in a collective.

What the freewill deniers usually ignore is that the brain is abundantly programmable by experiences and human choices. For those experiences that are not imposed, who does the choosing? In large measure, each of us uses our own consciousness to program what we believe, think, and do.

The conscious avatar that believes in its power to choose also must believe that it can exert some control over the automatic purposes of its unconscious mind. Such avatars believe they can train and discipline their minds, bending them to freely willed purposes. Such avatars say to themselves, "I can quit smoking," "I can make myself learn how to be an engineer," I will make this marriage work," "I will not lie, cheat, or steal, nor will I tolerate those who do," and so on.

Now it is true that all of us avatars commonly surrender our decision-making autonomy to our other two minds. Much of our behavior is unthinking, knee-jerk responsiveness. Our capacity for free will is limited to our willingness to claim and exert it. In other words, do we believe we have free will?

People who believe that humans have no free will and that consciousness has no causal actions are forced to rely on speculation rather than on sound research. Freewill deniers are hard-pressed to explain how people are not responsible for their choices and actions. What is it that compels foolish or deviant behavior? Is our avatar compelled to believe in God or to be an atheist? Is our avatar compelled to accept one moral code over other codes? Are we compelled to become a certain kind of person, with no option to "improve" in any self-determined way? Have we no free choice of learning experiences? If so, what or who does the compelling? Are we inevitable victims of genetics and experience or even a robotic unconscious mind?

If we believe that we can't make consciously reasoned choices, then there is not much we can do to improve ourselves or our plight in life. Or even if there are things that can be done to change us and our situations, the approach will surely have to be different if we can't initiate the change by force of our free will. The government or schools or religious doctrine or some other outside force must program our unconscious mind. Indeed, that is an increasingly popular position. Maybe most people, not just scholars, believe that free will is an illusion. That, of course, is a driving force behind moves to increase the size and power of government and federalize education at all levels. Powerful governments become tyrannical when enough people accept their dependency on government. Similarly, education may become indoctrination; religion may become so doctrinaire that it denies us permission to think for ourselves.

Believing there is no "I" in charge removes the need to demand or expect personal responsibility. All manner of bad brains and bad behavior can be excused—and thereby perpetuated.

THE AVATAR'S "NEW" GENETICS

The ability of the conscious avatar to *do* things becomes magnified in the light of the "new" genetics known as epigenetics. We used to think that who we are as conscious beings is determined by the genes we inherit. That is highly misleading. Genes, whose function is to code for proteins, only do things if they are expressed, and expression depends on the environment—including the mental environment.

When a protein-coding gene is expressed, its double-helix strands of DNA become unzipped to expose the coding nucleotides so they can translate the code into the RNA that will then translate the code into the various proteins used by the body. Most of the DNA previously thought to be "junk" has now been revealed to regulate gene expression by way of enhancing or suppressing expression of protein-coding genes.

A decade-long international research collaboration involving 442 scientists in 32 worldwide institutions now makes obsolete the original scientific views about DNA and genes. This research initiative, called ENCODE, was initiated and largely funded by the Genome Research Institute of the US National Institutes of Health. In just one week in November 2012, evidence for a new view was presented in some thirty groundbreaking publications in premier journals.[85]

Most of our genes do not code for the RNA that translates into proteins. The ENCODE project did confirm that there are about twenty-one thousand traditional protein-coding genes, but they constitute only about 3 percent of the human genome. Until now, all the other DNA was thought to be "junk DNA," presumably left over from ancient ancestors and with no function in today's evolved species.

ENCODE scientists have discovered that about 80 percent of the human genome does have function, but that DNA transcribes RNA as end products that regulate the expression of protein-coding genes. In short, there are two kinds of RNA, protein-coding RNA and regulatory RNA. This is really big news!

Over eighteen thousand species of regulatory RNA have been described. Clusters of regulatory genes make regulatory RNA occur throughout the chromosomes, and they very often regulate nonadjacent genes, often working with other regulatory genes as a team.

These new findings cause some scientists to assert a need to redefine what a gene is. The basic unit of heredity, they say, is not DNA, but rather its RNA transcripts. Why is this new view important? *Production of regulatory RNA is governed by the environment.* These "epigenetic" influences include things like what you eat, your bodily activities, what you think, and the feedback from how you behave. For example, a body builder, through intense exercise, causes expression of genes that create extraordinary muscle mass. Think of Arnold Schwarzenegger before, during, and after his body building career. He had the same protein-coding genes all along, but their expression changed by what he chose to do.

What this says is that *you* can control the expression of many of your genes by the choices you make and by what you think and do. Most of us are born with a genome that can generate a happy and productive life. Whether or not that happens depends on how our choices and actions affect gene expression.

So, next time you are tempted to blame your genes for a bad outcome, consider what role you played in the expression of those genes. Science is showing the truth of *Proverbs* 23:7),

As a man thinks in his heart, so is he.

CHAPTER 5
TO BEYOND?

Each individual life is a victory against the second law of thermodynamics. Life's adventure from womb to tomb is characterized by stemming the inexorable tide of maximum entropy (randomness), forcing molecules into orderly function against all odds. At death, the bodily molecules degrade, lose the ordering that made life possible, and the body's atoms return to random movement. "Dust to dust" is not a precise description of death, but it's not far off the mark.

Death at the tissue level typically occurs first in the brain, for during life the brain consumes a disproportionate share of the energy-producing oxygen needed to thwart the second law. In a matter a five minutes or so without oxygen, the brain dies. It can last a few minutes longer in children because their brains consume less oxygen than does the adult brain.

Throughout human history, all cultures have resisted accepting death as the ultimate end. Each culture has devised belief systems that allow for some kind of afterlife. Until recently, there was absolutely no scientific evidence to support any kind of afterlife. Indeed, it is well known that most scientists are either atheists or agnostics, although the numbers who believe in some kind of God and an afterlife are substantial.

In recent decades, two things have happened that make it a little harder for scientists to be so sure about their rejection of religion.

1. Modern physics.
2. Apparently valid reports of out-of-body experience in people who recover from clinically declared death.

I will discuss each in turn.

"SPOOKY" SCIENCE: HIDDEN REALITIES

Physics is the mother science. As such, it holds the greatest power for discovering the true nature of the universe and life within it. Physicists these days seem preoccupied with astronomical issues, such as the origin and ultimate fate of the universe. But some physicists venture into the realm of biology, claiming that their unique experimental and mathematical skills give them special insight into matters of life and death.

I just hate it when physicists write about biology. They sometimes say uninformed and silly things. But I hate it just as much when I write about physics, for I too am liable to say uninformed and silly things—as I may well do here.

To digress briefly, I am reminded of the communication gap between people of science and everybody else, as so powerfully discussed by C. P. Snow in his classic book *Two Cultures*. These days, within science there are also two cultures: physical science and biological science, and they don't always speak the same language. The language of physics, for example, relies heavily on mathematics, which is rarely mastered by biologists.

For most of my career, biology was generally considered a "soft" science, unworthy of the same stature as physics and chemistry. The discovery of DNA structure gave biology new respect in the "hard science" community because DNA is simple, as clearly explainable as chemistry, and easy to measure with mathematics. But the rest of biology is still a second-class science. I remember my College-of-Science dean, a nuclear physicist, refused to allow me to offer a course in sociobiology, based on E. O. Wilson's classic text, because the he did not consider such studies to be real science. He also objected to my publishing with experimental economists on the same grounds.

It's hard for biologists to argue with physicists. Often physicists listen with detached bemusement because biologists can't explain life with mathematics. Physics could not exist without math. Sometimes I think physicists get too enamored with math. I get the impression that they think that describing and predicting phenomena with equations is the same as explaining why and how such phenomena occur. Take the most famous equation of all, $E = mc^2$. Just what does that equal sign mean? It implies that the variables on each side are the same. But is mass really identical to energy? True, mass can be converted to energy, as atom bombs prove, and energy can even be turned into mass. Still, they are not the same things. Not only are the units of measurement different, but the equation is only descriptive and predictive. It does not explain *how* mass converts to energy or vice versa.

The limits of math become more troublesome when physicists try to explain the origin of the universe. Math does not really explain how a universe can exist without a first cause. True, physicists invoke the "big bang," a massive explosion of supercondensed matter. They call this the "singularity," as if that explains things any better. Whatever words, or math, they use, they cannot explain what created the supercondensed mass in the first place. Where did that mass come from? If it was created by energy, where did that come from? You can see that such questions create an infinite loop of effects that have a cause. Scientists call this "infinite regression," which is an untenable way to explain anything.

Even if you invoke the idea of a creator god, where did that god come from? So, you see, physicists and the rest of us are stuck with the unsatisfying conclusion that something can be created from nothing. I have only read one explanation for how this might happen, which I will discuss shortly, but it makes no sense to me.

Surely, many mysteries of the universe and of life itself are well hidden. Science is in the business of revealing hidden realities. What we call religious beliefs may be among those realities. Maybe we should revisit the view of the ancient Greek philosophers who held that there is "true" reality hidden by what we think is reality.

Today, physicists are starting to see previously unseen realities, as I am about to summarize. Such unseen realities may well include unknown kinds of matter and energy that give rise to mind. Maybe there is a counterpart mind, operating in parallel in a way that electrodes and amplifiers or magnetic imaging scanners cannot detect.

Only a few neuroscientists argue that the human mind is not materialistic. Neuroscientist Mario Beauregard and journalist Denyse O'Leary have written a whole book to argue the point.[1] Their *Spiritual Brain* documents many apparent mystical experiences. These authors use the existence of such mental phenomena as intuition, will power, and the medical placebo effect to argue that mind is spiritual, not material. None of this is proof that such experiences have no material basis. Their argument seems specious. They have no clear definition of *spirit*, and they do not explain how spirit can change neuronal activity or how neuronal activity translates into spirit. They dismiss the points I made earlier that mind can affect the brain because it originates in the brain and can modify and program neural processes because mind itself consists of neural process.

Sometimes we don't see hidden realities even when they are right under our nose. Consider water, for example, which before the advent of science was grossly misunderstood. Now we can explain how water exists in several states: liquid, vapor, solid. You and I are mostly water. My point is that our mental essence may also exist in several states. At the moment, the only one you and I know about is the state of nerve impulse patterns. Just as water has no way to *know* which state it is in, I (so far at least) can only know about my impulse-pattern state.

By now readers know brains make sense (pun intended). That is, we know enough about the brain to know that conscious mind may someday be explained by science. We already know enough about the nonconscious mind of the brain-stem and spinal cord to realize that what we call mind has a material basis that can be explained by science. Science may someday be able to examine what we today call spiritual matters. Consider the possibility that "spirit" is actually some physical property that scientists do not yet understand.

The idea of a material, biological basis of conscious mind may be offensive to those who believe in the mysteries of the soul and eternity. After all, many people of faith refuse to accept science's doctrine of evolution. To these believers we could say that one of the least mysterious ways God works in the world is through the laws of chemistry and physics that govern the universe and all living things. Even God has to have methods for doing things. Educated believers surely have to admit the possibility that God created these laws as a way to create the universe and even the human mind. Otherwise, from that perspective, what are the laws for? Nobody knows how these laws came to be or why they exist.

Many scientists are not sanguine about their belief in a material mind. For example, one scientist-engineer, Paul Nunez, has suggested that some yet-to-be-discovered information field might interact with brains such that brains act like a kind of "antenna," analogous to the way the retina of the eye can be thought of as an antenna that detects the part of the electromagnetic spectrum we call light.[2]

To me, other possibilities for discovering material attributes of "spirit" seem more likely. Modern physics, especially quantum mechanics and the theories of relativity, dark matter, and dark energy, has already shown that not even physicists understand what "material" is. I will now summarize the more likely possibilities for hidden realities of mind.

Quantum Mechanics (QM). Quantum mechanics is so weird that Einstein called it "spooky science."[3] Ironically, there remains a spooky weirdness in Einstein's own relativity theories, which I will get to momentarily.

The heart of the QM enigma lies in the apparent fact that subatomic particles can be in two places at the same time. But that is not quite correct. What has been demonstrated experimentally is that photons or electrons can have characteristics of both waves and particles at the same time. Where the wave and/or particle is located depends on whether or not its location is pinned down by observation. That observation includes instruments, not just the human eye.

Moreover, the waves are actually mathematical wave functions of the probability of where a particle is located. The shape of the probability of the wave function as it evolves can actually be quantified by the so-called Schrödinger equation. When we observe where a particle is located, the probability function "collapses," going from zero percent probability for all the locations where the part is not found to 100 percent for the place where it is observed.

But beyond the math, some particles, like photons, are clearly waves that oscillate at particular frequencies. The physics community was rocked in the 1920s by experiments that showed that electrons, known at the time to be subatomic particles, behaved like common waves, interfering with each other when their waves overlapped, much as two ripples in water do as the ripples move into each other. Electron interference seems to depend on a wave from one place crossing another wave from another place. How can that be? Max Born in 1927 found the answer: the waves are not physical waves but probability waves. Specifically, the size of a probability wave at any given point of location is proportional to the probability that the electron is located at that location. Stated in another way, the wave function tells us the probability of finding a particle at any given point of space. A profound consequence is that the probability wave applies to all locations in the universe.

Some of the experimentally demonstrable spooky things about QM include a seeming influence on elementary particles from distant parts of the universe with no time delay (called entanglement), particles jumping from one place to another without ever locating in places in between like successive frames in a motion picture (called tunneling), that particles can be in more than one place at the same time, and that the behavior of a particle is governed by its being observed or measured impersonally by instruments.

Knowing about QM is not the same as understanding it.[4] Even Heisenberg's uncertainty principle, a bedrock of QM theory, has recently been called into question.[5]

A key enigma in QM is that we can only observe a tiny subset of what actually exists. In QM theory, you can't make a complete observation, even remotely with instruments, of an object or event without disrupting its actual existence. The location of an object, for example, is one of several states: it may here or several places there. But in QM these states are specified as wave functions, not "here" or "there." Wave functions are probability statements. The object has, for example, a 75 percent chance of being in one place and a 25 percent change of being in another. Where it actually is depends on whether or not we detect its location. This is confusing I know, but I will let physicists do the apologizing.

To date, there is no compelling evidence that QM operates at levels beyond subatomic particles. But how can we be sure? QM might even be a basis for what we would otherwise think of as nonmaterial consciousness. Indeed, views on QM consciousness are published in scientific journals, and one journal is devoted exclusively to QM consciousness.

The most recent idea I have read is that Shannon's information theory lies at the heart of QM and can explain how something can emerge from nothing.[6] Information, quantified as "bits" (0 and 1) is inversely proportional to the probability of an occurrence (with probability measured on a logarithmic scale). I always wonder why physical scientists like to express things in inverse relationships. Anyway, the equation says that "information" has only two properties: an event and its probability of happening. The equation applies to any kind of event, from occurrences today to the moment the universe came into being. Moreover, the amount of information contained in an event is directly proportional to how unlikely it is to occur. Unlikely events do happen, and their rarity gives them the most information.

Physicist Vlatko Vedral, in his *Decoding Reality*, asserts that QM can resolve disputes over whether the world is random or deterministic. The enigma is that quantum events are random, but large objects behave deterministically (that is, are effects with causes). The key point is that quantum events can also be deterministic (quantified by the Schrödinger equation). For example, experiments using a beam-splitter mirror show that a photon can seem to be in two places at the same time (that is, that it has gone through the mirror and has also been reflected by it). But when you try to detect where the photon is, it will randomly appear in only one place (behind the mirror or in front of it). The mere act of observing, even if you do it indirectly with some kind of instrument, affects where the photon is. If that is not spooky, what is?

The corollary is that this science seems to suggest that we humans create reality by observation. This point of view is philosophical solipsism, which was championed by Walter Seegers in a book chapter he wrote for an earlier book of mine.[7] Seegers was a pioneer in the discovery of many of the mechanisms of blood clotting. Along the way, he came to the philosophical conclusion that science does not exist except in our own minds. He approvingly quotes Arthur Eddington, "We have found a strange footprint on the shores of the unknown. We have devised profound theories, one after another, to account for its origin. At last we have succeeded in reconstructing the creature that made the footprint. And Lo! It is our own." In the solipsistic view, the conscious sense of self discussed earlier now has a new dimension beyond developing events along the continuum of womb to tomb.

Vedral's view of reality is a little different. He has not explicitly integrated QM into solipsism, nor has anybody else as far as I know. But some of the ideas seem related. Vedral's main point is that random events can exist as a deterministic reality when they occur without being detected—as was likely the case at the birth of the universe when there was presumably nothing around to do the detecting. Today's reality is supposedly created by our observation, either directly or remotely via instrumentation.

QM gives a new dimension to information theory, for now quantification can be done in terms of "qubits," which can exist in multiple states as any combination of yes or no, on or off, and the like. This view of reality assumes the universe is digital. But my experience with biology, especially brain function, is that life is analog. Analog properties vary continuously, not as digital events of on or off. We use digital sampling and measurement of life events as a convenience. In fact, it is so convenient that we come to mistakenly believe that the world really is digital.

The first thing that qubits have to explain is the first law of thermodynamics, which says that energy—and by extension, matter—cannot be created from nothing. The universe supposedly arose from the big bang explosion of supercondensed matter. Where did that matter come from? To explain the inexplicable, Vedral speculates that subatomic particles exist only as the labels we use to describe the outcomes of experimental observations. He claims that "any particle of matter . . . is defined with respect to an intricate procedure that is used to detect it." If particles only exist in the presence of a detector, then the nothingness of the pre-universe developed a reality only when something that could detect a reality appeared. Sounds like gibberish to me. What was that first detector? Where and how did it appear?

There are multiple scholars who think consciousness may someday be explained by QM. With great trepidation, as a biologist suggesting to physicists how to study this matter, I would advise focusing on the wave function aspects of QM. Brain electrical currents, which are the currency of thought, still have magnetic properties, even though the current is carried by ions not electrons. There are sophisticated imaging devices that can monitor such magnetic fields, and they are used to produce a magnetoencephalogram.

Relativity. Einstein never came to grips with QM. I've had physicists tell me that had Einstein seen the evidence gathered since his death, he would surely have become a believer in what he had called "spooky science." Yet Einstein's own discoveries have their own spookiness. His theories have stood such a long test of time that some scientists are lured into thinking they understand relativity better than they actually do.

Most people know that Einstein discovered relativity. First, there was special relativity, which held that time is a fourth dimension that is relative depending on the location and speed of objects used as a frame of reference, that increasing speed of an object causes time to slow down, and that the only constant time is the speed of light. And of course there is the famous $E = mc^2$ equation that holds that mass and energy are interconvertible. Most of these seemingly wild ideas have been experimentally verified.

But nobody talks about the possible relevance of these ideas to brain function and consciousness. Of course, relativity effects are measurable only at high speeds. Does anything in the brain moves at high speeds? What about the propagation of voltage fields associated with nerve impulses? The brain does have a high-speed passive spread of voltage fields from multiple moving ionic currents. Also, what about the energy generated as electrons whip through protein chains in mitochondria? Only some of the energy is trapped in phosphate bonds of adenosine triphosphate. We *assume* that all the other energy is lost as heat. How can we be sure relativity is irrelevant to energy capture? Energy is well established as crucial for consciousness.

Many years later Einstein added variable movements and gravity to his theory to produce the general theory of relativity. In this perspective, time and space are wedded in an inseparable space-time continuum in which space is filled with the gravitational forces of stars and earths that cause space to bend and stretch "the fabric" of space-time. Think of space as a three-dimensional rubber sheet that is bent where bowling balls (stars and planets) occur within it. We think we know what this means on cosmic scales. What does it mean at the level of cells in the brain and the microgravity of their cellular mass and the time course of their chemical activities?

An added complication is that recent research confirms Einstein's original conjecture that gravity exists as ripples in the curvature of space-time that propagate as a wave, traveling outward from the source. Thus, we should think of gravity radiation as a form of energy release by objects with mass. There is a group at my own Texas A & M University actively engaged in study of such radiation.

But all studies of gravity radiation are done at the macro level of the universe. Does not our own body have mass? The molecules within our body have mass. Do they not have microgravity radiation? If so, what does such radiation do? Gravity waves oscillate, in theory at a variety of frequencies. Could this have anything to do with rhythms in the brain? Most scientists would probably discount such possibilities because gravity waves are so weak. But the ones we study, from distant galaxies, are weak because they are so far away. The mass in our body may emit weak gravity that is close at hand.

Moreover, think about the implications of general relativity's "continuum." That implies infinity. Our being and life locate on this space-time continuum. Maybe death is just one (temporary) point on the continuum.

String Theory. Physicists agree that relativity and quantum mechanics are in conflict, yet both theories stand on solid experimental ground. A major thrust of physics research today is devoted to finding how to reconcile these two views of the universe. String Theory is one of several mathematical approaches to resolving the conflicts. String theory holds that ultimate reality exists not as particles but as miniscule vibrating "strings" whose oscillations give rise to all the particles and energy in the universe and—nobody mentions—in our brain! The requirement for oscillation in vibrating strings should resonate with our emerging understanding of the role of oscillation in brain function and also with what was said above about gravity waves. What information is contained in the vibrating strings inside the atomic particles of neurons? Where did the vibrating strings come from? If string theory is correct, it will likely have great explanatory power for all living matter.

Parallel universes. Mathematically, string theory only works correctly if there are 11 dimensions or "universes." If there are such parallel universes, where are they "out there?" Some physicists imagine our universe like an expanding bubble inside a froth of space that is spawning multiple universe bubbles. Moreover, like foam in beer, each bubble might contain some portion of the properties of the parent source of froth.

Does the matter of our bodies simultaneously exist in more than one universe? Can bubbles in the froth of multiple universes interact, perhaps through quantum entanglement, or even coalesce? Perhaps what happens in our own inner universe of the brain is mirrored in another universe.

These esoteric ideas are gradually coming within the scope of experimental science. The new Large Hadron Collider particle accelerator on the Swiss-French border is designed to test string theory among other things. If the theory is correct, the collider should generate a host of exotic particles we never knew existed. One example is the Higgs boson, tentatively confirmed in 2013.

Another line of evidence might come from the Planck satellite to be launched by the European space satellite consortium. Some string-theory models predict that there is a specific geometry in space that will bend light in specific ways that the satellite is designed to detect.

String theory is not accepted by all physicists. But most agree that the known facts of physics do not fit any alternative unifying theory. Whatever theory emerges from accumulating evidence, it will, like Darwin's theory of evolution, revolutionize our thinking about the world and our life.

Dark matter. One of those parallel universes may be right under our nose. I'm talking about the massive amount of "dark matter," which astronomers believe to have mass because they see light being bent, presumably by gravity. This light bending occurs in regions of space where there is no observable matter to generate the gravitational force (see figure 5.1). This unseen matter is also inferred because it is the only known way to account for the rotational speed of galaxies, the orbital speed of galaxies in clusters, and the temperature distribution of hot gas in galaxies.

Last Spring, a fifteen-month census of the universe's matter by the European Space Agency calculated that this invisible matter accounts for 26.8 percent of the universe and that our ordinary matter accounts for only 4.6 percent. Everything else is energy.

Another thing to ponder: galaxies differ in their amount of dark matter, depending on the size of the galaxy.[8] The really interesting questions deal with possible interactions of dark matter and detectable matter. Are they totally independent? Or do they interact in some way we don't know about?

If dark matter is spread around the universe, and living things are created out of the matter of the universe, shouldn't some dark matter reside inside of us? Are properties of regular matter mirrored in dark matter? Is any part of us mirrored in dark matter? Similar questions could be asked about dark energy.

The false-color image of colliding galaxy clusters shows visible matter as red, dark matter as blue.

Figure 5.1. Dark matter in the universe. Black-and-white version of a false-color rendition of dark matter (hazy areas on the right and left of center) associated with colliding galaxy clusters. (Photo courtesy of NASA. X-ray: NASA/CXC/CfA/M. Markevitch et al.; Optical: NASA/STScl; Magellan/U. Arizona/D. Clowe et al.; Lensing Map: NASA/STScl; ESO WFI; Magellan/U. Arizona/D. Clowe et al.)

Dark energy. In 1998, two teams of researchers deduced from observing exploding stars that the universe is not only expanding but doing so at an accelerating rate. Forces of gravity should be slowing down expansion, and indeed they do seem to hold each galaxy together. But the galaxies are flying away from each other at incredible, accelerating speed.

Think of the "big bang" theory as a supercondensed hand grenade, which when it explodes sends shrapnel in all directions. The difference is that when the universe was born the pieces of its shrapnel (stars and planets, organized as galaxies) started accelerating as they moved apart.

The only sensible way to explain accelerating expansion is to invoke a form of energy, a "dark" energy that we don't otherwise know how to observe, that is pushing galaxies farther apart in a nonlinear way. Clearly, this dark energy is by far the most powerful force in the universe.

Why wouldn't some of that dark energy be within us? If so, it would obviously have to be present in relatively miniscule amounts, lest we blow up. All that we know is that energy has to be absorbed by its target to have any effect. When we get a sun burn, for example, enough of the sun's energy is absorbed in our skin to damage it. In the case of radiation, like x-rays and gamma rays, the absorption is ionizing: that is, electrons are knocked out of atoms as the energy is absorbed, leaving positive ions in the wake. An x-ray print shows the image created as a result of the rays that passed through your tissue hitting the photo-sensitive molecules in the film to darken them. Bone, for example, appears white because it is more likely to absorb x-rays and not allow them access to the photo-graphic plate. Gamma rays have much more energy, and when they are absorbed by tissue they can cause greater damage, even setting up DNA changes that can lead to cancer.

So what about dark energy? To push galaxies apart, it must impart some of its energy to the cluster of stars and planets to give them a push. What must dark energy be doing to us? Obviously, its push is not greater than the gravity that keeps us fixed to earth. But if that energy is absorbed by the galaxy, surely some of it must be absorbed in us. But what could such absorption do? Would such dark energy interact with the regular energy that we know about—like the energy in our brain? Could it act on consciousness?

There are still larger questions. Science is still trying to explain how ordinary matter and energy arose from the "big bang." Science does not even know how to start investigating where dark matter and dark energy came from.

NEAR-DEATH EXPERIENCES

In recent years a scientific and spiritual enigma has been created by multiple independent medical reports of out-of-body experiences in people who have recovered from being clinically dead. Such people after being resuscitated from cardiac arrest independently report a similar experience during their "death" of some combination of seeing bizarre visions of tunnels or bright lights, feeling themselves hovering over their body, and sensing overwhelming love.

In many such cases, the EEG was recorded as a vital-sign monitor during heart surgery, and death was indicated by arrested heart function and a flat EEG. When there is no EEG signal, there are no circuit impulse patterns. Without CIPs, there is no thought. So how could such a brain have such thoughts during the cardiac arrest? Though electrical brain waves were not always recorded during near-death experiences, it is a good bet that when clinical signs indicate this death, the brain waves have ceased. There is a possibility that the EEG recording parameters were not adequate. Pen-and-ink recordings cannot display much signal above thirty cycles per second. If computer monitored, the electronic filters may not have been set to allow amplification of signals in the high-frequency range that is characteristic of higher conscious thought. Nonetheless, it is commonly accepted that a flat-line EEG, even in the pen-and-ink cases, does correlate with ceased function of the neocortex, the crucial area that sustains consciousness.

As an aside, my student Joe Mikeska and I once recorded EEGs from rats after either cutting off their oxygen supply or decapitation to see if these common methods of lab-animal sacrifice were humane. We found that the EEG becomes flat on average around seventy seconds after asphyxiation and about thirteen seconds after decapitation.[9] This basic finding has since been confirmed by others. A human brain likely stops functioning much sooner because it needs disproportionate amounts of oxygen. So the long stories of near-death experience seem unlikely to be experienced in such short time. Moreover, all that we know about memory consolidation indicates that shutting down the brain immediately after a learning experience will prevent it from ever forming a memory of whatever happened just before the shutdown.

Also, consider this: I once took US Air Force altitude physiology training, in which the oxygen for jet pilots is turned off so that they can learn what that feels like and take corrective action. It's like being drunk, except that you go completely blank, I think in less than a minute. When the attendant sees this behavioral failure, oxygen is immediately provided and one's return to consciousness seems to switch on immediately. They had me write a poem as the oxygen was cut off. Mine was the lyric "Mary had a little lamb." When I checked my writing later, I didn't get past "it's fleece was white as . . . white as. . . ." To my knowledge, no fighter pilot trainee has ever reported anything like a near-death experience while their brain was shut down from oxygen deprivation.

You could argue correctly that a brain can last maybe four or five minutes without oxygen (and thus no neocortical CIPs). Nobody can explain how such a "dead" brain can create experiences of tunnels, bright lights, and so on, much less remember them.

The incidence of such experiences is not rare. Sam Parnia, a physician at New York Presbyterian Medical Center claims that about 10 percent of patients who recover from cardiac arrest report similar mental experiences while they are clinically dead.[10] A study in the Netherlands revealed that 18 percent of the 344 cardiac-arrest patients reported near-death experiences.[11]

Another research team studying near-death reports in Belgium reported that about 10 percent of the people in their survey had similar memories of near-death experiences that were exceptionally strong and vivid even years after the experience.[12] Irrespective of culture or religion, people in their survey reported similar memories of being out of body, passing through a tunnel, river, or door toward warm, glowing light, being greeted by dead loved ones, and being sent back to their bodies or told it's not time to go yet.

Scientists try to explain away such experiences as confabulated stories, but I have to ask why all these people independently have essentially the same story. Then there is the flimsy explanation that the dead brain is not really dead, that it has some undetected activity.

Though science cannot dismiss such reports, it cannot do much about investigating them either. Science has no theory and no tools to examine such phenomena. Science has no answers to the question of what happens at the end of life beyond the functions of apoptosis, autotrophic bacteria, and lysosomal enzymes that hasten the decay of flesh into the second law's disorder. What science does know makes it clear that life is more than the sum of its chemistry and physics. We are still in the dark about life and death.

LIVING THE LIFE WE WERE MEANT TO LIVE

Exotic theories of physics may someday explain how we all got here and what our brain's odyssey is all about. It's not likely that science will explain *why* we are here. In the absence of knowing for sure our purpose in life, it only makes practical sense that we fabricate our own purposes and at least take charge of those things over which we do have control.

Hopefully, this book has convinced you that you can and should be more proactive in sculpting your brain and who you are. You need not be an inevitable victim of your genes or your environment. The brain's capacity for learning and gene expression, the plasticity of its anatomy and physiological functions, and yes, a degree of free will, all provide the means for each person to be his or her own master. The corollary is that everyone is personally responsible for their choices and what they make of themselves—at least you can "be all you can be." What you think and do changes your brain and thus yourself as a person. People have expressed this philosophy in many ways.[13]

Always continue the climb. It is possible for you to do whatever you choose, if you first get to know who you are and are willing to work with a power that is greater than ourselves to do it.
> —Ella Wheeler Wilcox

God gave us the gift of life; it is up to us to give ourselves the gift of living well.
> —Voltaire

When we are no longer able to change a situation, we are challenged to change ourselves.
> —Viktor E. Frankl

It is not in the stars to hold our destiny but in ourselves.
> —William Shakespeare

No one saves us but ourselves. No one can and no one may. We ourselves must walk the path.
> —Buddha

Emancipate yourselves from mental slavery, none but ourselves can free our minds!
> —Bob Marley

In the long run, we shape our lives, and we shape ourselves. The process never ends until we die. And the choices we make are ultimately our own responsibility.

—Eleanor Roosevelt

The world is the great gymnasium where we come to make ourselves strong.

—Swami Vivekananda

If we did all the things we are capable of, we would literally astound ourselves.

—Thomas A. Edison

Good actions give strength to ourselves and inspire good actions in others.

—Plato

We must free ourselves of the hope that the sea will ever rest. We must learn to sail in high winds.

—Aristotle Onassis

Our ambition should be to rule ourselves, the true kingdom for each one of us; and true progress is to know more, and be more, and to do more.

—Oscar Wilde

There is nothing deep down inside us except what we have put there ourselves.

—Richard Rorty

Life is not easy for any of us. But what of that? We must have perseverance and above all confidence in ourselves. We must believe that we are gifted for something and that this thing must be attained.

—Marie Curie

Taking charge requires rational action. Science is revealing that irrational thinking often arises deep within the unconscious emotional centers of the brain. When we are emotional, areas in the emotion-controlling parts of the brain preferentially light up with more neural activity. Although the rational-thinking neocortex is still operative, such a fire blazing beneath it can overwhelm rational thinking.

Rational programming of attitudes and behavior demands that we some-times must delay gratification, such as studying hard about a topic that is boring but important, avoiding casual sex, leaving a cushion in the checking account, or investing for retirement. The behaviors are still driven by our brain's reward system, but our conscious rational mind has to teach the reward system that it is often worth waiting for your reward or that bad unintended consequences can accompany failure to wait. Programming of the brain is taken to a new level by the conscious, rational mind. It is as if the brain generates a conscious mind in order to maximize and optimize self-programming.

The conscious, rational mind should also decide which rewards serve our best interests. Smoking pot is certainly rewarding, but the rational mind can decide that it's wise not to risk screwing up brain function. Not only does common sense tell us that the conscious mind can identify the reinforcers that we should not pursue, but recent research shows that the decision-making parts of the brain are tightly coupled to the brain's reward system, thus allowing higher thought processes to program the reward system.

Brain programming, unlike computer programming, is seldom a one-step process. In the brain, programs are learned through repetition. Rational mind can dictate whether you pursue given attitudes, feelings, and behavior enough to make them become lasting parts of your program. It's a choice.

Most of us have heard the now-trite expression, "Life is what you make of it." My view has a different twist. What we make of life is what we make of ourselves. If you remember nothing else from this book, please remember that what you have chosen to hear, see, and think has sculpted changes in your brain that have in turn created what you have become as well as a bias for what you will become. Change of oneself is not only possible, but also obligatory if you wish to escape self-imposed limitations and become more successful and self-actualized. You can teach your brain to unlearn bad habits and thoughts that serve you poorly.

You can teach your brain better attitudes and ways of thinking. You can learn to change attitudes, emotions, beliefs, and behaviors that embarrass you and cause you to make excuses.

Conscious awareness is a great expediter for programming the brain. Ideally, the process operates like this: If you want to program your brain to change, you first have to be aware of how you stand at the moment. What is the "bad brain" problem that you want to change? Then, of course, you must make a conscious decision to make the change, typically weighing the costs versus the anticipated benefits. You must remind yourself that you have the free will to succeed. You can do this! Next, you must develop a plan, one based on "life-by-the-inch" small-step principles that will lead you step-by-step to programming the right changes. The plan should state what your problem is and list in order the steps you must take for deliverance. Write down the intrinsic benefits of each step as it is successfully completed, but in addition, write down other rewards you will give yourself along the way. Savor your successes and learn to do more of what enabled the success.

Give yourself appropriate rewards along the way. The results need to be monitored, providing feedback to your aware mental analyzer so you know how you are progressing and what adjustments, if any, need to be made. These are operations of the conscious mind, though they can be supported appropriately by timed rewards for your unconscious mind.

Perhaps the biggest problem people have in getting motivated to make a change in their lives is that they have not thought carefully enough about the price they are paying for remaining in their comfort zone. It is all too easy to rationalize that things are not so bad and that the benefits of change are not really that different. A rationalization is no more than a seemingly logical explanation to justify an excuse. Continued long enough, a rationalization eventually becomes a belief one has talked oneself into. And it becomes easier to believe in a rationalization as more barriers to making a change are erected. The remedy is, of course, to demand of yourself a clear-eyed look at who you are, where you are in your brain's odyssey, and how change might make you better off. Sometimes, if you are willing to listen, friends or relatives will tell you. Enemies may even make the message clearer. In the end, though, this is an analysis you have to make for yourself. No denial, no deception, no excuses. Just look for the road to deliverance and start hiking.

No longer do you have to make excuses or wallow in the self-pity of victimization. No longer do you have to be stuck in a career rut or be unhappy. No longer do you have to be paralyzed by fear, whether fear of failure or fear of success. Becoming your own agent of change is empowering. It liberates you.

For many people, it is not enough to have a program for needed change. For example, I have two colleagues at work who had open-heart surgery. Both lost about fifty pounds immediately after surgery, but within a year or so, both had gained back all the weight they had lost. Another colleague was morbidly obese and had part of his stomach removed to help him lose weight. There was no sign from looking at him two years later that he had lost any weight. These colleagues had programs to solve their weight problems. They knew how to change what they ate and how much they ate. They knew to chew slowly so that they could savor and feel satisfied with less food. They knew they had to exercise rigorously and consistently. All these program elements for change were in place. But they simply didn't run the program.

Running a life-change program can certainly be done. I have another colleague at work who had a mild heart attack. For him, it was a wake-up call that he answered. He lost about fifty pounds and has kept it off for the last ten years. The difference is that his mind ran his program.

PERSPECTIVES ON SCIENCE AND THE MIND'S ODYSSEY

Whatever the meaning of our life, it began when we, as fetuses, burst upon the world with eyes that had never seen light. Our brain's odyssey began in the dark. Presumably it ends that way—except for the unnerving fact that survivors of clinical death claim that brilliant light shines brightly again at death.

As the brain continues its growth and organization as a fetus and in the first years after birth, we acquire a conscious sense of self, operating I think as an avatar the brain has constructed to act on its behalf.

People say, "I came into this world alone, and I will leave the same way." But that's not exactly true. We all came in with our mother, certainly. Probably there was a doctor and a nurse or family member. Nonetheless, we were pushed alone into this world without our intent or permission. In all cases, we came in with some kind of life force that we scientists can only partially describe and certainly do not understand.

Our existence and emerging sense of self were marked from the embryonic gastrula to fetus, to expulsion through that dark birth canal. Our entrance into this world was marked by a tunnel of darkness, and then we discovered there was glowing light "out there." As we leave this world, it will become "dark in here" again, but maybe that, too, will yield to glowing light somewhere "out there." Maybe we will be greeted by open arms, as I was in the beginning.

As for what our brain's do during our lifetime, some science is misleading. Scientists' assertions that free will does not exist and that consciousness can't do anything thus far do not withstand scrutiny, as I think I showed in chapter 4.

Science does not advance by trying to test hypotheses that are stated in the negative. Examples would include: "Free will does not exist." "Consciousness can't do anything." "There is no creator God." The problem with such hypotheses is the classic logical difficulty of trying to prove a negative.

Scientists, at least astronomers and physicists, think they understand the "big bang" origin of the universe, but they can't explain the antecedent singularity that they say caused the "big bang," nor can they explain the creation of quarks, protons, and electrons, the weirdness of QM, the incompatibilities of QM and relativity, why there is inexplicable dark matter and dark energy, or life itself, particularly the essence of human consciousness. One should not worship equations. Likewise, I advise against worshipping holy books.

Surely, the fundamentals of nature are just too magical and miraculous to be the result strictly of random events. I think it is irrational to assert with any confidence that my life journey or yours are just the result of a long series of random accidents. Where are the experiments to show that? How can scientists accept on the one hand the laws of chemistry and physics as governing the universe and claim on the other hand that everything results from chance events? Biologists like to assert that evolution progresses from random genetic change. I think that is a careless use of the idea of randomness.

As mentioned, life cannot be sustained by randomness. Life exists because it thwarts the second law of thermodynamics, which holds that all matter must degrade into randomness and chaos. Death is inevitable, but life can hold it at bay for many years. Some trees, being less complex than humans, can live for hundreds of years.

And what about the first law of thermodynamics? It says that energy is neither created nor destroyed, it only transforms from one form to another. That means that energy cannot be created from nothing. The Big Bang theory holds that the energy of the universe came from the explosion of a supercondensed mass. What created that mass? How did it get so compressed that it had to blow up? New discoveries like that of the Higgs field are exciting, but they don't address the most basic questions. What kind of field is the Higgs boson? Where did it come from? Even if we had the right data, the evidence could only apply to ordinary matter, not the much more prevalent dark matter and dark energy.

Secular physicists claim that the original matter and energy of the universe arose out of nothing. What experiments have they designed to test that? Where is their evidence? They also have no experiments to show where the laws of chemistry and physics came from.

Science, for all its sophistication and progress, has not yet and may never answer the deepest questions of life. Kant's original contention that humans cannot directly know anything about true reality may need to be extended to include the limitations of the theories and instruments that scientists use to discover true reality. Maybe this is a fool's errand for science.

Even with the powerful tools and methods of science, our understanding is constrained by the limitations of the human mind. Just as my dog's little brain leaves her confused about many things she experiences, at the higher level of human mind, our brains are probably still too small to understand everything.

We scientists, brain scientists above all, need to be reminded of how much we don't know. Yet scientists need to be encouraged and enabled to purge the darkness of our understanding and shed new light on the true nature of reality.

At this moment, we cannot know with certainty when consciousness first arose in evolution or when it first appears in human infants. We cannot know for certain if consciousness is a circuit-impulse-pattern avatar. We cannot know the extent, if any, of free will or whether we are accountable for what we choose to believe and do. We cannot be sure of what it is that consciousness does. We cannot know if there is life after death.

Scholars of logic and philosophy have long insisted that supportive evidence alone is not sufficient to identify truth. Scientific truth is what is left after evidence proves what is untrue. Yes, a standard axiom of science is that absence of evidence is not evidence of absence.

A major existential issue persists. I made the point earlier that life seems to temporarily violate the second law of thermodynamics, delaying the inevitable destruction of life into chaos and randomness. But let us consider a scientific heresy: maybe the second law is wrong or at least incomplete. How can this law explain that:

(1) the universe is not collapsing in on itself, rather new solar systems continue being born, and the universe is expanding;

(2) the dark energy that drives the universe's expansion is not dissipating, in fact it has to be increasing to account for the acceleration;

(3) despite massive extinctions of plants and animals, total biomass may be increasing—certainly human population is;

(4) death is just a local perturbation of the second law, a prelude to another kind of living reality that does not end in randomness?

Science cannot resolve these matters, at least at present. But science confirms our intuition that these issues remain as valid possibilities. Even if we can't prove or disprove our beliefs, it does matter. What we believe about our brain's odyssey governs how we live out that odyssey.

It's dark in here. Let there be light!

NOTES

PREFACE

1. See A. Nguyen, "Why Rene Descartes' 'I Think, Therefore I Am' Does Not Prove Our Existence," Yahoo! Voices, http://voices.yahoo.com/why-rene-descartes-think-therefore-am -does-not-11592540.html (accessed July 9, 2013).

CHAPTER 1. IN THE BEGINNING

1. N. N. Taleb, *The Black Swan: The Impact of the Highly Improbable* (New York: Random House, 2010).

CHAPTER 2. HOW BRAINS WORK

1. G. M. Edelman, *The Remembered Present: A Biological Theory of Consciousness* (New York: Basic Books, 1989).

2. M. Scheffer et al, "Anticipating Critical Transitions," *Science* 338 (2012): 344–47.

3. G. M. Edelman, *Bright Air, Brilliant Fire: On the Matter of the Mind* (New York: Basic Books, 1992).

4. E. M. Macphail, *The Evolution of Consciousness* (New York: Oxford University Press, 1998).

5. W. R. Klemm, *Atoms of Mind* (New York: Springer, 2011).

6. M. Y. El-Naggar and S. E. Finkel, "Live Wires," *Scientist* (May 2013): 38–43.

7. A. Araque and M. Navarrete, "Glial Cells in Neuronal Network Function," *Philosophical Transactions of the Royal Society B* 365 (2010): 2375–81.

8. T. Binzegger et al., "Cortical Architecture," in M. DeGregorio et al., eds., *Brain, Vision, and Artificial Intelligence*, Lecture Notes in Computer Science 3704 (Berlin/Heidelberg: Springer, 2005).

9. Donald Stein, "Obama's Brain Map Initiative Needs a Rethinking," Livescience, http:// www.livescience.com/28505–map-the–brain.html (accessed July 8, 2013).

10. R. J. Douglas and K. A. Martin, "Neuronal Circuits of the Neocortex," *Annual Review of Neuroscience* 27 (2004): 419–51.

11. A. Burkhalter, "Many Specialists for Suppressing Cortical Excitation," *Frontiers in Neuroscience* 2 (2008): 155–67.

12. W. R. Klemm and C. J. Sherry, "Do Neurons Process Information by Relative Intervals in Spike Trains?" *Neuroscience and Biobehavioral Reviews* 6 (1982): 429–37.

13. A. Celletti and A. E. P. Villa, "Low-Dimensional Chaotic Attractors in the Rat Brain," *Biological Cybernetics* 74 (1996): 387–93.

14. K. Yun-Hui et al., "Repetition Rates of Specific Interval Patterns in Single Spike Train Reflect Excitation Level of Specific Receptor Types, Shown by High-Speed Favored-Pattern Detection Method," *Brain Research* 1113 (2006): 110–28.

15. G. Buzsáki and A. Draguhn, "Neuronal Oscillations in Cortical Networks," *Science* 304 (2004): 1926–29.

16. K. J. Miller, B. L. Foster, and C. J. Honey, "Does Rhythmic Entrainment Represent a Generalized Mechanism for Organizing Computation in the Brain?" *Frontiers in Computational Neuroscience* 6, no. 85 (2012), http://www.ncbi.nlm.nih.gov/pmc/articles/PMC3480650/ (accessed December 10, 2013).

17. J. H. Austin, *Zen Brain Reflections* (Cambridge, MA: MIT Press, 2006).

18. W. Singer, "Binding by Synchrony," *Scholarpedia* 2, no. 12 (2007): 1657.

19. W. R. Klemm, *Core Ideas in Neuroscience*, 2nd ed. (Bryan, TX: Benecton, 2013).

20. C. L. Striemer, S. Ferber, and J. Danckert, "Spatial Working Memory Deficits Represent a Core Challenge for Rehabilitating Neglect," *Frontiers in Human Neuroscience* 7 (2013), http://www.frontiersin.org/Journal/10.3389/fnhum.2013.00334/full (accessed December 10, 2013).

21. E. Mignot, "The Perfect Hypnotic," *Science* 340 (2013): 3638.

22. Y. I. Fishman, C. Micheyl, and M. Steinschneider, "Neural Representation of Harmonic Complex Tones in Primary Auditory Cortex of the Awake Monkey," *Journal of Neuroscience* 33, no. 25 (2013): 10312–10323.

23. M. Boly, "Preserved Feedforward but Impaired Top-Down Processes in the Vegetative State," *Science* 332 (2011): 858–62.

24. W. R. Klemm, "Sense of Self and Consciousness: Nature, Origins, Mechanisms, and Implications," in A. E. Cavanna and A. Nani, ed., *Consciousness: States, Mechanisms, and Disorders* (Hauppauge, NY: Nova Science, 2012), available at https://www.novapublishers.com/catalog/product_info.php?products_id=38801 (accessed December 10, 2013).

25. Karin Schwab et al., "Nonlinear Analysis and Modeling of Cortical Activation and Deactivation Patterns in the Immature Fetal Electrocorticogram," *Chaos* 19, no. 1 (2009).

26. O. Petre-Quadens, "Sleep in the Human Newborn," in O. Petre-Quadens and John D. Schlag, ed., *Basic Sleep Mechanisms* (New York: Academic Press, 1974), pp. 355–76.

27. D. Povinelli and S. Giambrone, "Reasoning about Beliefs: A Human Specialization?" *Child Development* 72 (2001): 691–95.

28. R. Perou et al., "Mental Health Surveillance among Children—United States, 2005–2011," *Morbidity and Mortality Weekly Report* 62, no. 2 (2013): 1–35, http://www.cdc.gov/mmwr/preview/mmwrhtml/su6202a1.htm (accessed July 9, 2013).

29. H. Harlow, "Birth of the Surrogate Mother," in W. R. Klemm, ed., *Discovery Processes in Modern Biology* (Huntington, NY: Kreiger Press, 1977), pp. 134, 150.

30. M. R. Rosenzweig, "Modification of Brain Circuits through Experience," in F. Bermúdez-Rattoni, ed., *Neural Plasticity and Memory* (Boca Raton, FL: CRC Press, 2007), pp. 67–94.

31. D. G. Stein, S. Brailowsky, and B. Will, *Brain Repair* (New York: Oxford University Press, 1995).

32. J. Freund et al., "Emergence of Individuality in Genetically Identical Mice," *Science* 340 (2013): 756–59.

CHAPTER 3. THE NATURE OF CONSCIOUSNESS

1. D. C. Dennett, *Sweet Dreams* (Cambridge, MA: MIT Press, 2005).

2. J. D. French and E. E. King, "Mechanisms Involved in the Anesthetic State," *Surgery* 38 (1955): 228–38.

3. J. F. Miller et al., "Neural Activity in Human Hippocampal Formation Reveals the Spatial Context of Retrieved Memories," *Science* 342 (2013): 1111–1114.

4. K. B. Swartz, "Self-Reflection, a Review of the Book, *The Face in the Mirror: The Search for the Origins of Consciousness*, by Julian Keenan with Gordon Gallup," *American Scientist* 91 (2003): 574–75.

5. V. Caggiano et al., "Mirror Neurons Differentially Encode the Peripersonal and Extrapersonal Space of Monkeys," *Science* 324 (2009): 403–406.

6. J. H. Kaas, "Topographic Maps Are Fundamental to Sensory Processing," *Brain Research Bulletin* 44 (1997): 107–112.

7. S. L. Craner and R. H. Ray, "Somatosensory Cortex of the Neonatal Pig: I. Topographic Organization of the Primary Somatosensory Cortex (SI)," *Journal of Comparative Neurology* 306, no. 1 (1991): 24–38.

8. W. R. Klemm, "Behavioral Inhibition," in W. R. Klemm and R. P. Vertes, ed., *Brainstem Mechanisms of Behavior* (New York: Wiley & Sons, 1990), pp. 497–533.

9. G. G. Gallup Jr., "Chimpanzees: Self Recognition," *Science* 167 (1970): 86–87.

10. C. Trevarthen, "Commissurotomy and Consciousness," in T. Bayne et al., ed., *The Oxford Companion to Consciousness* (New York: Oxford, 2009), pp. 158–63.

11. W. R. Klemm, "Behavioral Readiness," in W. R. Klemm and R. P. Vertes, ed., *Brainstem Mechanisms of Behavior* (New York: Wiley & Sons, 1990), pp. 105–45.

12. W. R. Klemm and R. P. Vertes, ed., *Brainstem Mechanisms of Behavior* (New York: Wiley & Sons, 1990).

13. C. Koch, *The Quest for Consciousness: A Neurobiological Approach* (Englewood, CO: Roberts, 2004).

14. W. R. Klemm, T. H. Li, and J. L. Hernandez, "Coherent EEG Indicators of Cognitive Binding during Ambiguous Figure Tasks," *Consciousness and Cognition* 9 (2000): 66–85.

15. R. Srinivasan, S. Thorpe, and P. L. Nunez, "Top-Down Influences on Local Networks: Basic Theory with Experimental Implications," *Frontiers in Computational Neuroscience* 7, no. 29 (April 2013), http://www.frontiersin.org/Journal/10.3389/fncom.2013.00029/full (accessed December 20, 2013).

16. T. Klingberg, *The Overflowing Brain: Information Overload and the Limits of Working Memory* (New York: Oxford University Press, 2009).

17. Joseph LeDoux, *Synaptic Self: How Our Brains Become Who We Are* (New York: Viking, 2009).

18. Y. Shrager et al., "Working Memory and the Organization of Brain Systems," *Journal of Neuroscience* 28, no. 18 (2008): 4818–4822.

19. S. M. Jaeggi et al., "Improving Fluid Intelligence with Training on Working Memory," *Proceedings of the National Academy of Sciences* 105, no. 19 (2008): 6829–6833, http://www.pnas.org/cgi/doi/10.1073/pnas.0801268105 (accessed December 20, 2013).

20. N. Cowan, *Working Memory Capacity* (New York: Taylor and Francis Group, 2005).

21. W. R. Klemm, *Memory Power 101* (New York: Skyhorse, 2012).

22. M. D. Melnick, B. R. Harrison, S. Park, et al., "A Strong Interactive Link between Sensory Discriminations and Intelligence," *Current Biology* 23 (2013): 1–5.

23. Klemm, *Memory Power 101*.

24. K. Takehara-Nishiuchi and B. L. McNaughton, "Spontaneous Changes of Neocortical Code for Associative Memory during Consolidation," *Science* 322 (2008): 960–63.

25. F. Kasanetz et al., "Transition to Addiction Is Associated with a Persistent Impairment in Synaptic Plasticity," *Science* 328 (2010): 1709–1712.

26. Julian Jaynes, *The Origin of Consciousness in the Breakdown of the Bicameral Mind* (Boston: Houghton Mifflin, 1976).

27. E. L. Merrin, T. C. Floyd, and G. Fein, "EEG Coherence in Unmedicated Schizophrenic Patients," *Biological Psychiatry* 25, no. 1 (1989): 60–66.

28. M. Massimini, F. Ferrarelli, R. Huber, et al., "Breakdown of Cortical Effective Connectivity during Sleep," *Science* 309 (2005): 2228–2232.

29. S. S. Dalal et al., "Intrinsic Coupling between Gamma Oscillations, Neuronal Discharges, and Slow Cortical Oscillations during Human Slow-wave Sleep," *Journal of Neuroscience* 30, no. 4 (2010): 14285–14287.

30. Emily Underwood, "Sleep: The Brain's Housekeeper?" *Science* 342 (2013): 301.

31. M. Bellesi et al., "Effects of Sleep and Wake on Oligodendrocytes and Their Precursors," *Journal of Neuroscience* 33, no. 36 (2013): 14288–14300.

32. Klemm, *Memory Power 101*.

33. S. Yoo et al., "A Deficit in the Ability to Form New Human Memories without Sleep," *Nature Neuroscience* 10 (2007): 385–92.

34. E. M. Robertson, D. Z. Press, and A. Pascual-Leone, "Off-Line Learning and the Primary Motor Cortex," *Journal of Neuroscience* 25, no. 27 (2005):6372–6378.

35. M. Czisch, T. C. Wette, C. Kaufmann, et al., "Altered Processing of Acoustic Stimuli during Sleep: Reduced Auditory Activation and Visual Deactivation Detected by a Combined fMRI/EEG Study," *Neuroimage* 16, no. 1 (2002): 251–58.

36. A. E. Prudom and W. R. Klemm, "Electrographic Correlates of Sleep Behavior in a Primitive Mammal, the Armadillo Dasypus novemcinctus," *Physiology and Behavior* 10 (1973): 275–82.

37. J. A. Hobson and R. McCarley, "The Brain as a Dream State Generator: An Activation-Synthesis Hypothesis of the Dream Process," *American Journal of Psychiatry* 134 (1977): 1335–1348.

38. W. R. Klemm, "Sleep and Paradoxical Sleep in Ruminants," *Proceedings of the Society for Experimental Biology and Medicine* 121 (1966): 635–38.

39. W. R. Klemm, "The Behavioral Readiness Response," in W. R. Klemm and R. P. Vertes, ed., *Brainstem Mechanisms of Behavior* (New York: John Wiley and Sons, 1990).

40. R. P. Vertes, "A Life-Sustaining Function for REM Sleep: A Theory," *Neuroscience and Biobehavioral Reviews* 10 (1986): 371–76.

41. W. B. Webb and H. W. Agnew Jr., "Stage 4 Sleep: Influence of Time Course Variables," *Science* 174 (1971): 1354–1356.

42. W. R. Klemm, "Why Does REM Sleep Occur? A Wake-up Hypothesis," *Frontiers in Neuroscience* 5, no. 73 (2011): 1–12.

43. J. M. Walz et al., "Simultaneous EEG-fMRI Reveals Temporal Evolution of Coupling between Supramodal cortical Attention Networks and the Brainstem," *Journal of Neuroscience* 33, no. 49 (203): 19212–19222.

44. C. Lauer et al., "A Polysomnographic Study on Drug-Naive Patients," *Neuropsychopharmacology* 16 (1997): 51–60.

CHAPTER 4. DOES CONSCIOUSNESS *DO* ANYTHING?

1. M. S. Gazzaniga, *The Mind's Past* (Berkeley: University of California Press, 1998).

2. E. O. Wilson, *Consilience: The Unity of Knowledge* (New York: Vintage Books, 1999), p. 137.

3. N. Di Pisapia, "Unconscious Information Processing in Executive Control," *Frontiers in Human Neuroscience*, January 31, 2013, http://www.fronticrsin.org/Human_Neuroscience/10.3389/fnhum.2013.00021/full (accessed December 26, 2013).

4. M. T. Banich, K. L. Mackiewicz, B. E. Depue, et al., "Cognitive Control Mechanisms, Emotions, and Memory: A Neural Perspective with Implications for Psychopathology," *Neuroscience and Biobehavioral Reviews* 33 (2009): 613–30.

5, A. T. Beck, "The Evolution of the Cognitive Model of Depression and Its Neurobiological Correlates," *American Journal of Psychiatry* 165 (2008): 969–77.

6. As quoted by Francis Crick in his presentation, "The Impact of Linus Pauling on Molecular Biology," Pauling Symposium, Salk Institute, Oregon State University, 1995, http://oregonstate.edu/dept/Special_Collections/subpages/ahp/1995symposium/crick.html (accessed December 30, 2013).

7. Jeff Dyer, Hal Gregersen, and C. M. Christensen, *The Innovator's DNA: Mastering the Five Skills of Disruptive Innovators* (Boston: Harvard Business Review Press, 2011).

8. P. Haggard, S. Clark, and J. Kalogeras, "Voluntary Action and Conscious Awareness," *Nature Neuroscience* 5 (2002): 382–85.

9. D. M. Wegner, "Who Is the Controller of Controlled Processes?" in R. Hassin et al., ed., *The New Unconscious* (Oxford: Oxford University Press, 2005).

10. S. Pockett, W. P. Banks, and S. Gallagher, ed., *Does Consciousness Cause Behavior?* (Cambridge, MA: MIT Press, 2009).

11. Gazzaniga, *Mind's Past*.

12. W. R. Klemm, "Neural Representations of the Sense of Self," Archives Cognitive Psychology, *Advances in Cognitive Psychology* 7 (2011): 16–30, http://www.ncbi.nlm.nih.gov/pmc/articles/PMC3163487/ (accessed December 27, 2013).

13. W. R. Klemm, "Are There EEG Correlates of Animal Thinking and Feeling?" *Neuropsychobiology* 26 (1993): 151–65.

14. R. Schulz et al., "Amnesia of the Epileptic Aura," *Neurology* 45, no. 2 (1995): 231–35.

15. J. Grafman and E. Wassermann, "Transcranial Magnetic Stimulation Can Measure and Modulate Learning and Memory," *Neuropsychologia* 37, no. 2 (1998): 159–67.

16. W. R. Klemm, "Typical Electroencephalograms: Vertebrates," in P. L. Altman and D. S. Dittmer, ed., *Biology Data Book*, vol. 2, 2nd ed. (Bethesda, MD: Federation of American Societies for Experimental Biology, 1973).

17. N. Wade, *The Faith Instinct: How Religion Evolved and Why It Endures* (New York: Penguin Press, 2009).

18. M. Egnor, "If Neuroscience Is a Victory for Materialism, What Would Defeat Look Like?" Evolution News and Views, December 11, 2008, http://www.evolutionnews.org/2008/12/we_can_relax_dr_novella_declar.html (accessed July 9, 2013).

19. R. J. Weber, *The Created Self* (New York: Norton, 2000).

20. P. E. Dux, J. G. Ivanoff, C. L. Asplund, and R. Marois, "Isolation of a Central Bottleneck of Information Processing with Time-Resolved fMRI," *Neuron* 52, no. 6 (2007): 1109–1120.

21. A. Newberg and M. R. Waldman, *Why We Believe What We Believe* (New York: Free Press, 2006).

22. E. O. Wilson, quoted in "E. O. Wilson," *Philosopedia*, http://philosopedia.org/index.php?title=E._O._Wilson (accessed December 30, 2013).

23. S. McMains and S. Kastner, "Interactions of Top-Down and Bottom-Up Mechanisms in Human Visual Cortex," *Journal of Neuroscience* 31, no. 2 (2011): 587–97.

24. B. A. Purcell, R. P. Heitz, J. Y. Cohen, et al., "Neurally Constrained Modeling of Perceptual Decision Making," *Psychological Review* 117, no. 4 (2012): 1113–1143.

25. A. N. McCoy and M. L. Platt, "Expectations and Outcomes: Decision-Making in the Primate Brain," *Journal of Comparative Physiology A* 191 (2005): 201–211.

26. Tim Allen, *I'm Not Really Here* (New York: Hyperion Books, 1996).

27. K. Kagaya and M. Takahata, "Readiness Discharge for Spontaneous Initiation of Walking in Crayfish," *Journal of Neuroscience* 30, no. 4 (2010): 1348–1362.

28. E. Warrington and L. Weiskrantz, "A Study of Learning and Retention in Amnesic Patients," *Neuropsychologia* 6, no. 3 (1968): 283–91.

29. Julian Jaynes, *The Origin of Consciousness in the Breakdown of the Bicameral Mind* (Boston: Houghton Mifflin, 1976).

30. H. Walter, *Neurophilosophy of Free Will: From Libertarian Illusions to a Concept of Natural Autonomy* (Cambridge, MA: MIT Press, 2001).

31. M. Jeannerod, "Consciousness of Action as an Embodied Consciousness," in S. Pockett, W. P. Banks, and S. Gallagher, ed., *Does Consciousness Cause Behavior?* (Cambridge, MA: MIT Press, 2009), pp. 25–38.

32. D. M. Wegner, *The Illusion of Conscious Will* (Cambridge, MA: MIT Press, 2002).

33. P. S. Churchland, "Self-Representation in Nervous Systems," *Science* 296 (2002): 308–10.

34. Gazzaniga, *Mind's Past*.

35. S. Pockett, W. P. Banks, and S. Gallagher, ed., *Does Consciousness Cause Behavior?* (Cambridge, MA: MIT Press, 2009).

36. W. R. Klemm, "Free Will Debates: Simple Experiments Are Not So Simple," *Advances in Cognitive Psychology* 6, no. 6 (2010): 47–65.

37. T. Bayne, "Phenomenology and the Feeling of Doing: Wegner on the Conscious Will," in S. Pockett, W. P. Banks, and S. Gallaher, ed., *Does Consciousness Cause Behavior?* (Cambridge, MA: MIT Press, 2009), pp. 169–85.

38. B. Libet and commentators, "Non-Conscious Cerebral Initiative and the Role of Conscious Will in Voluntary Action," *Behavioral and Brain Sciences* 8 (1985): 529–66.

39. H. C. Lau, R. D. Rogers, P. Haggard, and R. E. Passingham, "Attention to Intention," *Science* 303 (2004):1208–1210.

40. S. S. Obhi and P. Haggard, "Free Will and Free Won't," *American Scientist* 92 (2004): 358–65.

41. H. C. Lau, R. D. Rogers, and R. E. Passingham, "On Measuring the Perceived Onsets of Spontaneous Actions," *Journal of Neuroscience* 26 (2006): 7265–7271.

42. C. S. Soon et al., "Unconscious Determinants of Free Decisions in the Human Brain," *Nature Neuroscience* 11 (2008): 543–45.

43. T. Christophel and J. D. Haynes, "Single Trial Time-Frequency Decoding of Early Choice Related EEG Signals: Further Evidence for Non-Conscious Determinants of 'Free' Decisions," Program No. 194.19, Neuroscience Meeting Planner, Society for Neuroscience, Chicago, IL, 2009.

44. M. Desmurget et al., "Movement Intention after Parietal Cortex Stimulation in Humans," *Science* 324 (2009): 811–13.

45. J. Trevena and J. Miller, "Brain Preparation before a Voluntary Action: Evidence against Unconscious Movement Initiation," *Consciousness and Cognition* 19 (2010): 447–56.

46. H. G. Jo et al., "Spontaneous EEG Fluctuations Determine the Readiness Potential: Is Preconscious Brain Activation a Preparation Process to Move?" *Experimental Brain Research* 231, no. 4 (2013): 495–500.

47. A. Schurger et al., "An Accumulator Model for Spontaneous Neural Activity Prior to Self-Initiated Movement," *Proceedings of the National Academy of Sciences* 109, no. 42 (2012): E2904–E2913.

48. K. Grill-Spector and N. Kanwisher, "Visual Recognition: As Soon as You Know It Is There, You Know What It Is," *Psychological Science* 16 (2005): 152–60.

49. A. R. Mele, "Free Will: Theories, Analysis, and Data," in S. Pockett, W. P. Banks, and S. Gallagher, ed., *Does Consciousness Cause Behavior?* (Cambridge, MA: MIT Press, 2009), pp. 187–205.

50. F. Ono and J. Kawahara, "The Effect of Non-Conscious Priming on Temporal Production," *Consciousness and Cognition* 14 (2005): 474–82.

51. R. Ulrich, J. Nitschke, and T. Rammsayer, "Perceived Duration of Expected and Unexpected Stimuli," *Psychological Research* 70 (2006): 77–87.

52. S. Klein, "Libet's Temporal Anomalies: A Reassessment of the Data," *Consciousness and Cognition* 11 (2002): 198–214.

53. R. Nijhawan and K. Kirschfeld, "Analogous Mechanisms Compensate for Neural Delays in the Sensory and the Motor Pathways," *Current Biology* 13 (2003): 749–53.

54. S. Joordens et al., "When Timing the Mind One Should also Mind the Timing: Biases in the Measurement of Voluntary Actions," *Consciousness and Cognition* 11 (2002): 231–40.

55. A. N. Danquah et al., "Biases in the Subjective Timing of Perceptual Events: Libet et al. (1983) Revisited," *Consciousness and Cognition* 17 (2008): 616–27.

56. J.-C. Sarrazin et al., "How Do We Know What We Are Doing? Time, Intention, and Awareness of Action," *Consciousness and Cognition* 17 (2008): 602–15.

57. Desmurget et al., "Movement Intention after Parietal Cortex Stimulation."

58. S. Musallam, B. D. Cornell, B. Greger, et al., "Cognitive Control Signals for Neural Prosthetics," *Science* 305 (2004): 258–62.

59. W. R. Klemm, T. H. Li, and J. L. Hernandez, "Coherent EEG Indicators of Cognitive Binding during Ambiguous Figure Tasks," *Consciousness and Cognition* 9 (2000): 66–85.

60. S. Makeig, T. P. Jung, and T. T. Sejnowski, "Multiple Coherent Oscillatory Components of the Human Electroencephalogram (EEG) Differentially Modulated by Cognitive Events," *Society for Neuroscience Abstracts* 24 (1998): 507.

61. D. Lee, "Behavioral Context and Coherent Oscillations in the Supplementary Motor Area," *Journal of Neuroscience* 24 (2004): 4453–4459.

62. S. Van Gaal et al., "Frontal Cortex Mediates Non-Consciously Triggered Inhibitory Control," *Journal of Neuroscience* 28 (2008): 8053–8062.

63. N. N. Taleb, *The Black Swan: The Impact of the Highly Improbable*, 2nd ed. (New York: Random House, 2010).

64. T. Sommers, "The Illusion of Freedom Evolves," in Don Ross et al., ed., *Distributed Cognition and the Will* (Cambridge, MA: MIT Press, 2007), p. 73.

65. A. Schlegel et al., "Network Structure and Dynamics of the Mental Workspace," *Proceedings of the National Academy of Sciences*, September 2013.

66. K. D. Vohs and J. Schooler, "The Value of Believing in Free Will: Encouraging a Belief in Determinism Increases Cheating," *Psychological Science* 19 (2008): 49–54.

67. R. F. Baumeister, E. J. Masicampo, and C. N. DeWall, "Prosocial Benefits of Feeling Free: Disbelief in Free Will Increases Aggression and Reduces Helpfulness," *Personality and Social Psychology Bulletin* 35 (2009): 260–68.

68. T. F. Stillman et al., "Personal Philosophy and Personal Achievement: Belief in Free Will Predicts Better Job Performance," *Social Psychological and Personality Science* 1 (2010): 43–50.

69. M. E. P. Seligman, "Learned Helplessness," *Annual Review of Medicine* 23 (1972): 407–12.

70. B. Lee, *The Power Principle* (New York: Simon and Schuster, 1998).

71. A. Bandura, *Self-Efficacy: The Exercise of Control* (New York: W. H. Freeman, 1997).

72. R. Kane, "Agency, Responsibility, and Indeterminism: Reflections on Libertarian Theories of Free Will," in J. Campbell, M. O'Rourke, and D. Shier, ed., *Freedom and Determination* (Cambridge, MA: MIT Press, 2004), pp. 70–88.

73. N. Wade, *The Faith Instinct: How Religion Evolved and Why It Endures* (New York: Penguin, 2009), p. 243.

74. Disparate impact on housing was among the causes of the 2008 housing bubble burst (see M. Aleo and P. Svirsky, "Foreclosure Fallout: The Banking Industry's Attack on Disparate Impact Race Discrimination Claims under the Fair Housing Act and Equal Credit Opportunity Act," *Boston University Public Interest Law Journal* 1, Fall 2008). See also the Flood Insurance Reform Act of 2012, FEMA, http://www.fema.gov/flood-insurance-reform-act-2012 (accessed January 6, 2013); "Key Features of the Affordable Care Act," HealthCare.gov, http://www.hhs .gov/healthcare/facts/timeline/index.html (accessed January 6, 2013). For a discussion of disproportionate impact, refer to Title VII of the Civil Rights Act (see *EEOC v. Sambo's of Georgia, Inc.*, 530 F. Supp. 86, 92 [N.D. Ga. 1981]). For a discussion of the fairness doctrine, see D. Fletcher, "A Brief History of the Fairness Doctrine," *Time*, February 20, 2009, http://content.time.com/time/ nation/article/0,8599,1880786,00.html (accessed January 6, 2013).

75. L. Tancredi, *Hardwired Behavior: What Neuroscience Reveals about Morality* (New York: Cambridge University Press, 2005), pp. 46–61, 65–66.

76. Z. Torey, *The Crucible of Consciousness: An Integrated Theory of Mind and Brain* (Cambridge, MA: MIT Press, 2009).

77. D. Simons, "How Many Times Do They Pass the Ball?" [video], Smithsonianmag.com, http://www.smithsonianmag.com/multimedia/videos/How-Many-Times-Do-They-Pass-the -Ball .html (accessed January 3, 2014).

78. T. M. Pearce and D. W. Moran, "Strategy-Dependent Encoding of Planned Arm Movements in the Dorsal Premotor Cortex," *Science* 337 (2012): 984–88.

79. G. Mandler, "Consciousness: Respectable, Useful, and Probably Necessary," in B. J. Baars et al., ed., *Essential Sources in the Scientific Study of Consciousness* (Cambridge, MA: MIT Press, 2003), pp. 15, 33.

80. P. Tse, *The Neural Basis of Free Will: Criterial Causation* (Cambridge, MA: MIT Press, 2013).

81. Patricia Churchland, *Neurophilosophy: Toward a Unified Science of the Mind/Brain* (Cambridge, MA: MIT Press, 1986).

82. J. Shreeve, "Beyond the Brain," *National Geographic*, 2005, http://science.national geographic.com/science/health-and-human-body/human-body/mind-brain/ (accessed July 9, 2013).

83. W. R. Klemm, *Blame Game: How to Win It* (Bryan, TX: Benecton Press, 2008).

84. B. Lipton, *The Biology of Belief* (Santa Barbara, CA: Mountain of Love/Elite Books, 2005).

85. E. Pennizi, "ENCODE Project Writes Eulogy for Junk DNA," *Science* 337 (2012): 1159–1161.

CHAPTER 5. TO BEYOND?

1. M. Beauregard and D. O'Leary, *The Spiritual Brain* (New York: Harper One, 2007).

2. P. Nunez, *Brain, Mind, and the Structure of Reality* (New York: Oxford University Press, 2010).

3. A. Einstein, quoted in B. Greene, *The Fabric of the Cosmos: Space, Time, and the Texture of Reality* (New York: Vintage, 2004).

4. S. Carroll, *From Eternity to Here: The Quest for the Ultimate Theory of Time* (New York: Dutton, 2010).

5. J. Z. Salvail et al., "Full Characterization of Polarization States of Light Via Direct Measurement," *Nature Photonics* 7 (2013): 316–21.

6. V. Vedral, *Decoding Reality: The Universe as Quantum Information* (New York: Oxford University Press, 2010).

7. W. H. Seegers, "My Individual Science," in W. R. Klemm, ed., *Discovery Processes in Modern Biology* (Huntington, NY: Kreiger, 1977), pp. 242–60.

8. Y. Bhattacharjee, "Inventory Asks: Where Is All the *Non*-dark Matter Hiding?" *Science* 327 (2010): 258.

9. J. A. Mikeska and W. R. Klemm, "EEG Evaluation of Humaneness of Asphyxia and Decapitation Euthanasia of the Laboratory Rat," *Laboratory Animal Science* 25 (1975): 175–79.

10. S. Parnia, *What Happens When We Die* (Carlsbad, CA: Princess Books, Hay House, 2006).

11. P. van Lommel et al., "Near-Death Experience in Survivors of Cardiac Arrest: A Prospective Study in the Netherlands," *Lancet* 358, no. 9298 (2001): 2039–2045.

12. M. Thonnard et al., "Characteristics of Near-Death Experiences Memories as Compared to Real and Imagined Events Memories," *PLOS ONE* 8, no. 3 (2013): e57620.

13. The quotations that follow can be found at BrainyQuote, http://www.brainyquote .com/quotes/keywords/ourselves.html (accessed July 9, 2013).

INDEX

ABOUT THE AUTHOR

Early in his career Bill Klemm developed an interest in brain function from studying the nervous system while obtaining his Doctor of Veterinary Medicine degree from Auburn University, where he has since been honored as a distinguished alumnus. He then served briefly in the US Air Force and continued as a reservist in the Vision Research Laboratory and later in the Development Planning Office of the Human Systems Command, rising to the rank of colonel. He continued his neuroscience education and obtained a doctorate from the University of Notre Dame.

His first academic position was in the College of Veterinary Medicine at Iowa State University, where he was awarded tenure after his first year. Two years later he accepted an untenured position in the College of Science at Texas A & M University, where he again was awarded tenure after one year. Currently, he is Senior Professor of Neuroscience in the College of Veterinary Medicine.

Bill Klemm is a distinguished member of Sigma Xi, the Scientific Research Society. He has served on the official editorial boards of twelve scholarly journals, and editors of forty-five research journals have asked him to review approximately one thousand manuscripts. He has authored over two hundred peer-reviewed scholarly papers on a wide spectrum of neuroscience topics ranging from molecular models of membrane function to human cognition. He has written fifty-one book chapters and has authored seventeen books.

He is widely known for his expertise in science education, and his contributions to science education include the development of numerous neuroscience graduate and undergraduate courses. He was a pioneer in the development of Internet-based collaborative learning environments, cocreating a software program called FORUM that was a forerunner of Internet "wikis." He has authored much of the science curriculum material for middle schools at the website for Texas A & M's Partnership for Environmental Education and Rural Health.

Each month, he writes a newspaper column for baby boomers about age-related cognitive decline that is based on studies in the medical and neuroscience literature. He maintains a blog called *Improve Your Learning and Memory* and he also writes a blog for *Psychology Today*, whose editors have tagged many of his posts as "essential reads." His science blog posts have drawn nearly one million reader views.

In recent years, he has been listed in successive editions of *Who's Who in America* and *Who's Who in the World*. He lives in Bryan, Texas.